Creep Characteristics of Engineering Materials

Edited by Tomasz Tański, Marek Sroka, Adam Zieliński and Grzegorz Golański

Published in London, United Kingdom

IntechOpen

Supporting open minds since 2005

Creep Characteristics of Engineering Materials
http://dx.doi.org/10.5772/intechopen.78498
Edited by Tomasz Tański, Marek Sroka, Adam Zieliński and Grzegorz Golański

Contributors
Daniel C, Hemalatha G, Sarala L, Tensing D, Sundar Manoharan S, Amir Golalipour, Stan Mandziej, Tomasz Arkadiusz Tański, Adam Zieliński, Marek Sroka, Grzegorz Golański, Janusz Dobrzanski

First published in London, United Kingdom, 2020 by IntechOpen
IntechOpen is the global imprint of INTECHOPEN LIMITED, registered in England and Wales, registration number: 11086078, 7th floor, 10 Lower Thames Street, London, EC3R 6AF, United Kingdom
Printed in Croatia

British Library Cataloguing-in-Publication Data
A catalogue record for this book is available from the British Library

Additional hard and PDF copies can be obtained from orders@intechopen.com

Creep Characteristics of Engineering Materials
Edited by Tomasz Tański, Marek Sroka, Adam Zieliński and Grzegorz Golański
p. cm.
Print ISBN 978-1-78984-363-7
Online ISBN 978-1-78984-364-4
eBook (PDF) ISBN 978-1-78985-365-0

For EU product safety concerns:
IN TECH d.o.o., Prolaz Marije Krucifikse Kozulić 3, 51000 Rijeka, Croatia, info@intechopen.com or visit our website at intechopen.com.

Meet the editors

Prof. Tomasz Tański is the Head of the Institute of Engineering Materials and Biomaterials, Silesian University of Technology and a Member of the Metallurgy Section of the Committee of Metallurgy of the Polish Academy of Sciences. He is a specialist in non-ferrous alloys, composite materials, nanostructured manufacturing engineering, and surfaces. His interests include the study the properties and structures, and engineering materials. He has authored or co-authored more than 300 scientific publications worldwide, including 12 monographs and books, and more than 70 publications in the Philadelphia list. He has won 18 awards and honors both national and international. He has been both a supervisor and contractor for more than 15 research and didactic projects in Poland and abroad. He is also a reviewer and promoter of numerous scientific papers, including four doctoral research papers in the field of nanotechnology and materials.

Dr. Marek Sroka, PhD and MSc in Engineering, is an assistant professor at the Institute of Engineering Materials and Biomaterials at the Silesian University of Technology in Gliwice, Poland. During his scientific activity, he participated in and organized many scientific international conferences. His scientific interests include materials science and materials for service at elevated temperatures, in high-temperature creep resistance, in creep tests, and in computer-assisted material engineering. He is the author and coauthor of about 85 scientific publications worldwide, including more than 28 publications in the Philadelphia list. He has won 17 awards and honors, national and international, and has been a contractor for more than 10 research and didactic projects in Poland and abroad. He is also a reviewer of numerous scientific publications.

Prof. Adam Zieliński is a researcher at the Institute for Ferrous Metallurgy and an expert in the area of materials engineering. He is well versed in materials for service at elevated temperatures, in high-temperature creep resistance, in creep tests, and in the diagnostics of high-pressure power, chemical, and petrochemical installations. He collaborates closely with power boiler manufacturers, as well as the power industry repair plants in the area of power installation diagnostics, residual life, and damage processes. He is the author and coauthor of about 135 scientific publications worldwide, including more than 30 publications in the Philadelphia list, as well as author and coauthor more than 400 research works. He is also an expert on the direct instructions of the energy industry and petrochemical industry. He has won 10 awards.

Grzegorz Golański, PhD, is Professor at the Institute of Materials Engineering, Czestochowa University of Technology. He specializes in the study of structure, heat treatment, and properties of engineering materials, mainly creep-resistant alloys. He is the author and coauthor of about 230 scientific publications worldwide, including five monographs and books, and more than 30 publications in the Philadelphia list. He has won 15 awards. He is also the author and coauthor of about 90 research works and gives expert opinions for industries.

Contents

Preface

This book contains chapters with original and innovative research studies in the issues related to the broadly defined creep effect, which concerns not only the area of construction materials but also natural phenomena. In classical terms, creep is defined as a process that takes place under constant load, which ensures that stresses are usually within the Hooke's law range, and at elevated temperature. The loads applied to products working under creep conditions cause stresses and strains whose values may change over time, not only as a result of overloads occurring during service, but mainly at the time when loading remains unchanged and result in stresses lower than the yield strength.

In the creep of materials, both plastic and reversible strain occurs. As a result of the process, their shapes and dimensions change constantly during service. After the elapse of sufficient time of working under creep conditions, the components are degraded and their further service is prevented. The degradation processes start at the moment when the rate of strain changes from its minimum and proportional values to disproportional ones as it is increasing.

The development of materials for operation under creep conditions and the related manufacture of products with improved performance are a result of the dynamic worldwide demand for new technologies, and recently also those related to environmental protection. The result of the above-mentioned conditions is the extensive investigations of the properties of new materials carried out by scientists and researchers all over the world. This mainly concerns the components working under the so-called creep conditions, which are characterised by a slow change in the shape of material due to the operation of mainly prolonged permanent stresses, which are lower than the elastic limit of material.

The aim of next book from the series is to provide readers, students and PhD students, as well as research personnel and professional engineers involved in the operation of devices with comprehensive information on their changes in performance during work. Additionally, collection in one place of not only studies the phenomenon of creep but also the practical application of the discussed methods on specific examples and technological solutions. The chapters of this book will be developed by renowned and respected researchers and specialists from around the world.

Tomasz Tański and Marek Sroka
Silesian University of Technology,
Poland

Adam Zieliński
Institute for Ferrous Metallurgy,
Poland

Grzegorz Golański
Czestochowa University of Technology,
Poland

Chapter 1

Introductory Chapter: Creep - An Overview of New Research Results

Adam Zieliński, Marek Sroka, Tomasz Tański and Grzegorz Golański

1. Introduction

In classical terms, creep is defined as a process that takes place under constant load, which ensures that stresses are usually within the Hooke's law range, and at elevated temperature. The loads applied to products working under creep conditions cause stresses and strains whose values may change over time, not only as a result of overloads occurring during service, but mainly at the time when loading remains unchanged and results in stresses lower than the yield strength [1].

A characteristic feature of creep is that after strain relief of the material in service, the strain decreases and the so-called strain recovery takes place [2].

In the creep of materials, both plastic and reversible strains occur. As a result of the process, their shapes and dimensions change constantly during service. After the elapse of sufficient time of working under creep conditions, the components are degraded and their further service is prevented [3–5]. The degradation processes start at the moment when the rate of strain changes from its minimum and proportional values to disproportional ones as it is increasing [6].

The service within the time as assumed by designers as well as the extension of the lifetime of the materials and products made from them which have worked the design service life are most often determined by the calculation method based on creep strength data and positive results of comprehensive mechanical tests [7–11]. Among materials, the products operating under creep conditions are of particular importance. In their evaluation, the condition assessment of their materials is important and necessary [12–16]. It is carried out using destructive and non-destructive materials testing the result of which is referred to the available characteristics of materials both in the as-received condition and after service [17–22]. The results of these tests, supported by the extensive material characteristics database, allow for good estimation of the material condition and degree of exhaustion and determination of the time of further trouble-free service until the next scheduled inspection.

However, in a number of cases, to reach the design service life for products working under creep conditions, not only good estimation of their service life but also its determination based on destructive tests in a selected representative test area is required.

Yet, it is not always possible in practice. Therefore, the aspect of knowledge (experience), frequently acquired during long-term scientific and industrial practice, and also supported by thorough cost-effectiveness calculation made for such a procedure, is also important.

The problem in evaluation of creep resistance of materials operated under creep conditions for a long time is the time necessary to perform creep tests for

IntechOpen

the assessment of performance within their design service life and frequently also of the residual life, i.e., the remaining time of service beyond the design service life.

2. Outline of this book

This study presents the results of the research into the creep effect in different materials (ceramics, metallic materials, polymers and organic materials) and presents the method for using the assessment based on creep tests and numerical calculations to determine the actual lifetime. The publication also shows how to use the accelerated creep tests whose duration is reduced by changing one of the parameters in relation to the actual ones for such an assessment.

The development of materials for operation under creep conditions and the related manufacture of products with improved performance are a result of the dynamic worldwide demand for new technologies, and recently also those related to environmental protection. The result of the above-mentioned conditions is the extensive investigations of the properties of new materials carried out by scientists and researchers all over the world. This mainly concerns the components working under the so-called creep conditions, which are characterised by a slow change in the shape of material due to the operation of mainly prolonged permanent stresses, which are lower than the elastic limit of material.

The first part of the book entitled "Creep" concerned the creep effect and research methods applied for materials working under creep conditions. The publication consists of 14 interesting chapters developed by well-known and respected researchers and specialists from different countries. The editors received a lot of positive signals and feedback about high interest in the first part of the book in the scientific community. They were given numerous suggestions to continue the work and not only gather the issues related to the creep effect in a single study but also present the information on the practical application of the discussed methods based on specific examples and technological solutions. This subject has relevance as a significant development of new materials in which the creep effect is a decisive factor for their durability within the design service life has been observed in recent years. Therefore, there is a great demand for knowledge of the actual performance of materials during and beyond the design service life. When continuing their work on the next part of the book, the editors were guided by the words of the great Pole, Stanislaw Staszic: "Science and skills become useful only when they are applied in practice for the public use."

To recapitulate, the book aims to provide readers, including but not limited to MSc and PhD students as well as research personnel and engineers involved in operation of power equipment, with the comprehensive information on changes in the performance of creep-resistant materials during service.

Author details

Adam Zieliński[1], Marek Sroka[2*], Tomasz Tański[2] and Grzegorz Golański[3]

1 Institute for Ferrous Metallurgy, Gliwice, Poland

2 Institute of Engineering Materials and Biomaterials, Silesian University of Technology, Gliwice, Poland

3 Institute of Materials Engineering, Czestochowa University of Technology, Częstochowa, Poland

*Address all correspondence to: marek.sroka@polsl.pl

References

[1] Wójcik Z. Stanislaw Staszic's concept of science. Dissertations for the History of Education. 2006;**45**:85-100

[2] Zieliński A, Dobrzański J, Purzyńska H, Golański G. Properties, structure and creep resistance of austenitic steel Super 304H. Materials Testing. 2015;**57**:859-865

[3] Dziuba-Kałuża M, Zieliński A, Dobrzański J, Sroka M, Urbańczyk P. Residual life of boiler pressure parts made of the 13CrMo4-5 steel after long-term operation in a creep conditions. Archives of Metallurgy and Materials. 2018;**63**(2):889-897

[4] Golanski G, Zielinska-Lipiec A, Zielinski A, Sroka M. Effect of long-term service on microstructure and mechanical properties of martensitic 9% Cr Steel. Journal of Materials Engineering and Performance. 2017;**26**(3):1101-1107

[5] Sroka M, Zieliński A, Dziuba-Kałuża M, Kremzer M, Macek M, Jasiński A. Assessment of the residual life of steam pipeline material beyond the computational working time. Meta. 2017;**7**(3):82

[6] Zieliński A, Golański G, Dobrzański J, Sroka M. Creep resistance of VM12 steel. Archives of Metallurgy and Materials. 2016;**61**:1289-1294

[7] Sroka M, Zieliński A, Mikuła J. The service life of the repair welded joint of Cr-Mo/Cr-Mo-V. Archives of Metallurgy and Materials. 2016;**61**:969-974

[8] Golański G, Zieliński A, Zielińska-Lipiec A. Degradation of microstructure and mechanical properties in martensitic cast steel after ageing. Materialwissenschaft und Werkstofftechnik. 2015;**46**(3):248-255

[9] Sroka M, Zielinski A, Hernas A, Kania Z, Rozmus R, Tanski T, et al. The effect of long-term impact of elevated temperature on changes in the microstructure of Inconel 740H alloy. Metal. 2017;**56**(3-4):333-336

[10] Dudziak T, Deodeshmukh V, Backert L, Sobczak N, Witkowska M, Ratuszek W, et al. Phase investigations under steam oxidation process at 800°C for 1000 h of advanced steels and Ni-based alloys. Oxidation of Metals. 2017;**87**(1-2):139-158

[11] Zieliński A, Sroka M, Dudziak T. Microstructure and mechanical properties of Inconel 740H after long-term service. Materials. 2018;**11**:2130

[12] Purzyńska H, Golański G, Zieliński A, Dobrzański J, Sroka M. Precipitation study in Ti-stabilised austenitic stainless steel after 207,000 h of service. Materials at High Temperatures. 2019. DOI: 10.1080/09603409.2018.1546919

[13] Zieliński A, Dobrzański J, Purzyńska H, Sikora R, Dziuba-Kałuża M, Kania Z. Evaluation of creep strength of heterogeneous welded joint in HR6W alloy and Sanicro 25 steel. Archives of Metallurgy and Materials. 2017;**62**(4):2057-2064

[14] Sroka M, Nabialek M, Szota M, Zielinski A. The influence of the temperature and ageing time on the NiCr23Co12Mo alloy microstructure. Revista de Chimie. 2017;**68**(4):737-741

[15] Zieliński A, Sroka M, Miczka M, Śliwa A. Forecasting the particle diameter size distribution in P92 (X10CrWMoVNb9-2) steel after long-term ageing at 600 and 650°C. Archives of Metallurgy and Materials. 2016;**61**:753-760

[16] Zieliński A, Golański G, Sroka M. Influence of long-term ageing on

the microstructure and mechanical properties of T24 steel. Materials Science & Engineering, A: Structural Materials: Properties, Microstructure and Processing. 2017;**682**:664-672

[17] Zieliński A, Miczka M, Golański G. Forecasting the distribution of precipitate diameters in the presence of changes in the structure of the material. Archives of Metallurgy and Materials. 2017;**62**:283-290

[18] Golański G, Jasak J, Zieliński A, Kolan C, Urzynicok M, Wieczorek P. Quantitative analysis of stability of 9%Cr steel microstructure after long-term ageing. Archives of Metallurgy and Materials. 2017;**62**(1):273-281

[19] Zieliński A, Golański G, Sroka M, Skupień P. Microstructure and mechanical properties of the T23 steel after long-term ageing at elevated temperature. Materials at High Temperatures. 2016;**33**:154-163

[20] Kępa J, Golański G, Zieliński A, Brodziak-Hyska A. Precipitation process in VM12 steel after ageing at 650°C temperature. Journal of Vibroengineering. 2012;**59**(4):143-150

[21] Golański G, Lis A, Słania J, Zieliński A. Microstructural aspect of long term service of the austenitic TP347HFG steel. Archives of Metallurgy and Materials. 2015;**60**:2901-2904

[22] Zieliński A, Boryczko B, Miczka M, Sroka M. Forecasting in the presence of microstructural changes for the case of P91 steel after long-term ageing. Archives of Civil and Mechanical Engineering. 2016;**16**:813-824

Chapter 2

Effects of Creep on RC Frame Subjected to Cyclic Load with Magnetorheological Damper

Daniel Cruze, Hemalatha Gladston, Sarala Loganathan, Tensing Dharmaraj and Sundar Manoharan Solomon

Abstract

The study aims to discuss about a magnetorheological (MR) damper subjected to cyclic load test in reinforced concrete frames (RC). Two RC frames were cast, the dimension and detailing were adopted as per codal provisions. The effects of stress, strain, displacement and load behavior for RC element with various time interval is studied with and without MR damper. The typical creep curve of RC frame reveals, the creep rate decreases with time until reaching the steady state, after the initial deflection. The result shows that the incorporation of the MR damper reduced the displacement by 49% and an increase in load carrying capacity of 55% was attained compared with the RC frame without damper.

Keywords: RC frames, cyclic behavior, dynamic response, creep

1. Introduction

Semi active magnetorheological (MR) damper have a note of attention in the use of control systems to control vibrations in structures during earthquake. There are various semi active control systems which attracted much attention now a days because it possesses the adaptability of active control system. This MR damper makes the viscosity to more apparent, operate in less current, which produce high damping force. In semi active the properties will be varied dynamically, therefore which is quite effective in structural control responses due to its wide range loading condition [1–5]. The main advantage of semi active control device is their low power requirement. Particularly, the most promising device for seismic control is MR, MR fluid is the major constituent in MR damper which results controllable damper [6–9]. The characteristics of MR fluid is when applying magnetic field, their ability to change from liquid to semi solid state with the reaction time of less than 50 milliseconds. A typical MR fluid consists of 25–55% by volume of magnetic iron particles like carbonyl iron, suspended in an appropriate carrier liquid. The diameter of the particle was found to be 3–5 μm in size [10–16]. Furthermore, MR fluid is advantageous due to its unique phase transition property in the presence of magnetic field. However, the major disadvantage faced by MR fluid is sedimentation. Commonly, commercial lubricants were added to reduce gravity settling, which additionally enhance particle suspension [16–20].

The focus of this paper is to experimentally demonstrate the ability of the MR damper to reduce structural responses. Following a description of the experimental setup, cyclic load test is carried out in the RC frame with and without damper. The effects of stress, strain, displacement and load behavior for RC element with various time interval is studied with and without MR damper. The results indicate for seismic response reduction, MR damper is quite effective for wide range.

2. MR fluid

Magnetorheological fluid also called as MR fluid consist of magnetic material and a carrier oil influence effective under the applied magnetic field. The applied magnetic field have influence over most important rheological properties like as yield stress and apparent viscosity as shown in **Figure 1**. Their increased contribution in space technology, through domestic products is noteworthy. Particularly their applications in dampers, power steering pumps, brakes and control valves are significant. In automobile application, micron size particles are acceptable of the frequent vibrations but on move to seismic factor, since the vibrations are not regularly taken place, sedimentation cause major problems. Therefore, many researchers notably civil engineers are currently focusing on fluid behavior in dampers for structures. In the present work, MR fluid was prepared with the help of Nano Fe_3O_4 particle and silicone oil by 60% weight, additionally their efficiency towards MR damper is investigated.

2.1 X-ray diffraction studies

The synthesized nano Fe_3O_4 was well characterized by X-ray crystallographic technique and are depicted in **Figure 2** using the Model (Schimadzu, LAB X,

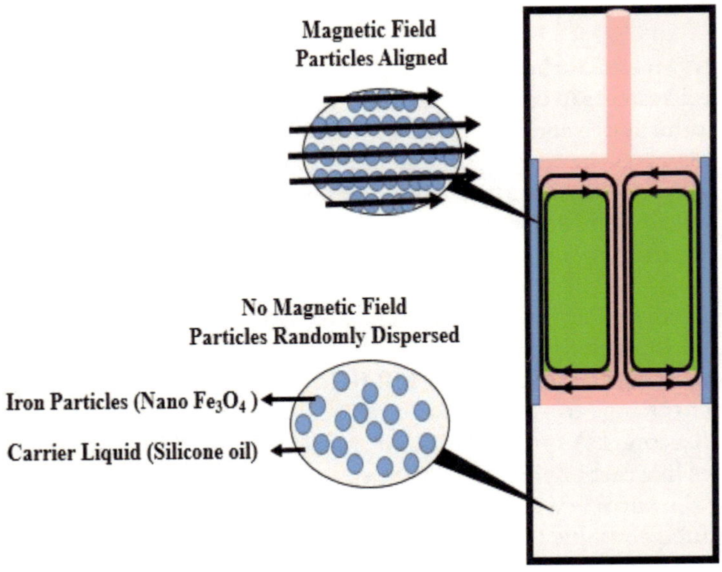

Figure 1.
MR damper with and without magnetic field.

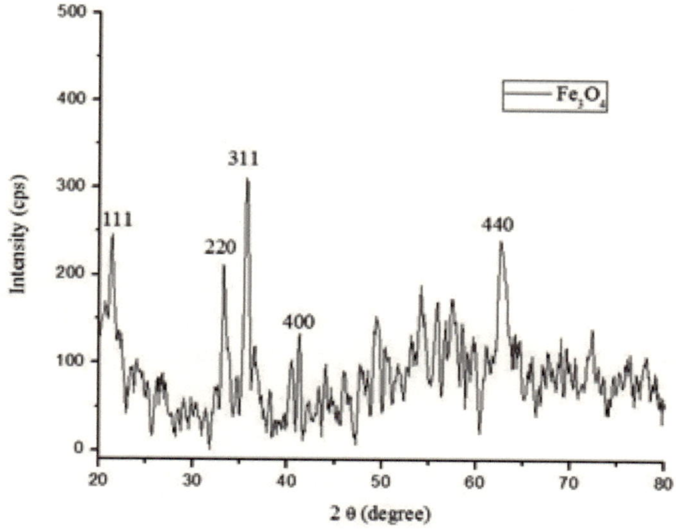

Figure 2.
XRD pattern of Fe_3O_4 iron particles.

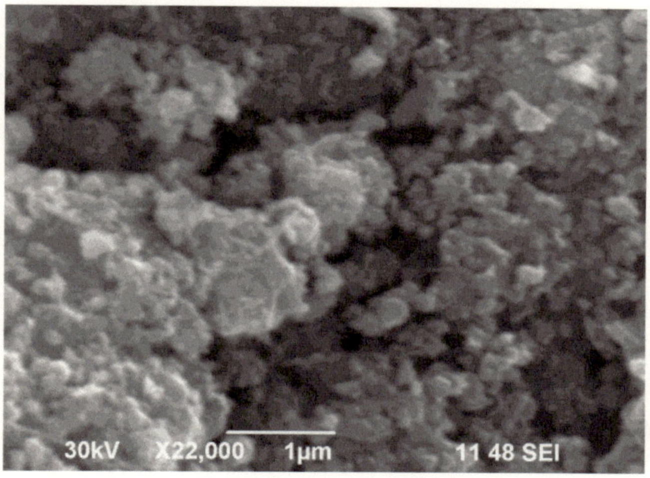

Figure 3.
SEM image of Fe_3O_4 iron particles.

XRD-600) using CuKα radiation. The prominent XRD peaks observed were indexed and analyzed using ICDD database. The JCPDS file no. 79-0417 represents a magnetite structure in it. The average particle size of the material was examined using the most intense plane (311) with the help of Scherrer equation. The average particle size is 12 nm.

2.2 Surface morphology studies

The surface morphology of the synthesized Fe_3O_4 was obtained using Scanning Electron Microscopy (SEM, JEOL 6390). **Figure 3** depicts the SEM image of the synthesized compound reveals the uneven spherical like structure for the Nano Fe_3O_4.

3. Experimental studies

3.1 Preparation of MR fluids

The nano sized iron particles (Fe_3O_4) of size 10–12 nm were used as a smart material and silicone oil as the carrier oil. The MR fluid was prepared by mixing 80% of weight of silicone oil and with 20% Fe_3O_4 particles. The addition of Fe_3O_4 was slow and uniform under stirring condition for about 30 minutes. Then, the above mixture is sonicated about 1 hour under the frequency of 0.5 Hz. During sonication, the major proportion dispersed with the silicon oil. Therefore, there discourages the gravity settling. The off-state viscosity of the MR fluid prepared at 27^0c is 0.822 Pa-s. **Figure 4** depicts the schematic representation for the preparation of MR fluid.

3.2 Modeling of MR damper

For the present study, the shear mode MR damper was modeled. **Figure 5a** shows the design and configuration of the MR damper adopted. The fabricated MR damper using mild steel was shown MR damper was fabricated using mild steel shown in **Figure 5b**. Since the proposed MR damper was shear mode, the copper coil was wound around the piston with 260 no. of turns.

From **Figure 5c**, the reasons behind the MR damper under shear mode is clearly explained. In the shear mode MR damper, a thin layer of MR fluid is sheared between the piston and the cylinder, that space is specifically called as annular gap. In order to produce large force, shear mode MR damper is needed.

3.3 Cyclic load test

The dimensions and the detailing of RCC frame adopted for the present work are given in **Figure 6**. The frame was casted according to the details given in **Figure 7**. The damper was scaled to resist forces in a single bay RC frame. The damper was positioned diagonally on the frame. The grade of concrete used for the specimen is M25, The reinforcement details of beam, column and stirrups of 8,8,3 mm. The support condition is fixed for the specimen.

3.4 Test setup

With the help of servo hydraulic actuator, in-plane lateral load was supplied to the specimen at the end of the top beam along its centroidal axis. The opposite side was supported by a strong steel reaction frame two channel section were placed symmetrically on both side of the beam and fitted tightly using strong bolts at regular intervals. The lateral load supplied to the beam using the channel section. In order to avoid the distraction particularly horizontal & vertical movements during

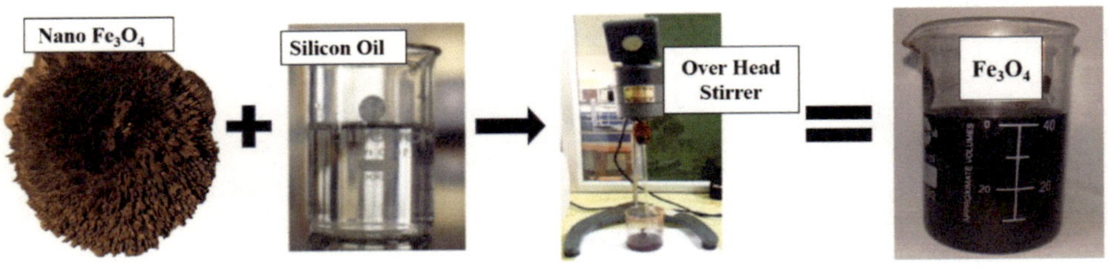

Figure 4.
Preparation of MR fluid.

lateral loading, the footing was fixed firmly in the strong floor. The lateral support was given by the steel sections to avoid it's out of plane movements. The test setup was given in **Figure 8**.

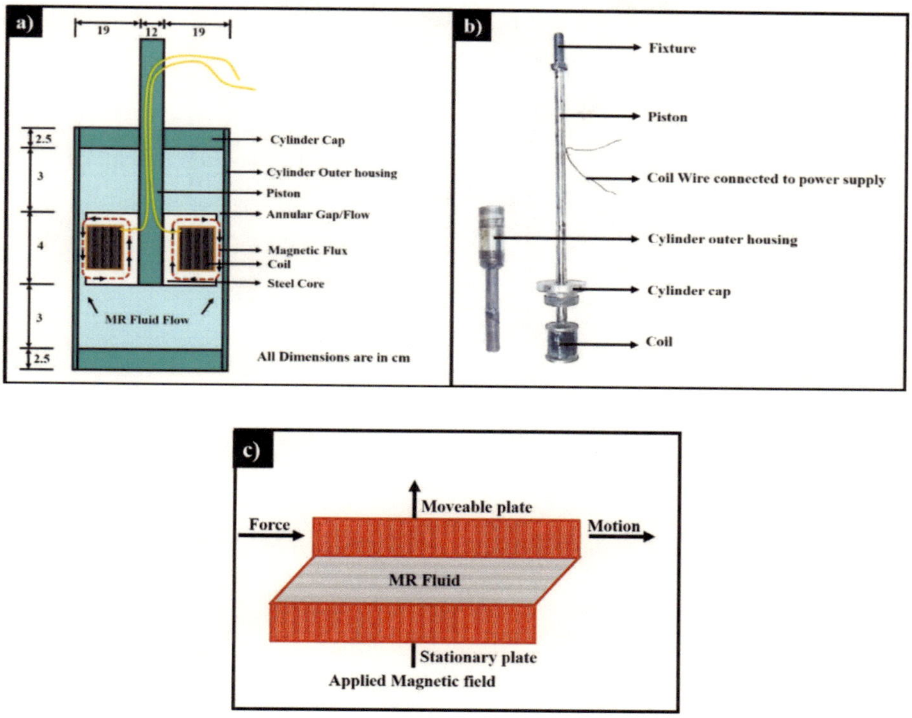

Figure 5.
(a) Schematic MR damper; (b) fabricated MR damper; (c) shear mode.

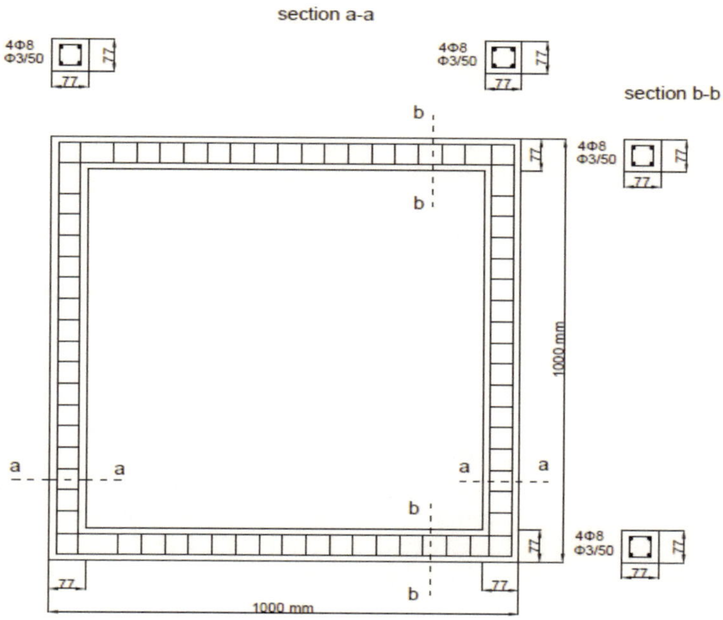

Figure 6.
Typical sketch of RC frame specimen detailing.

Figure 7.
Casted RC framed specimen.

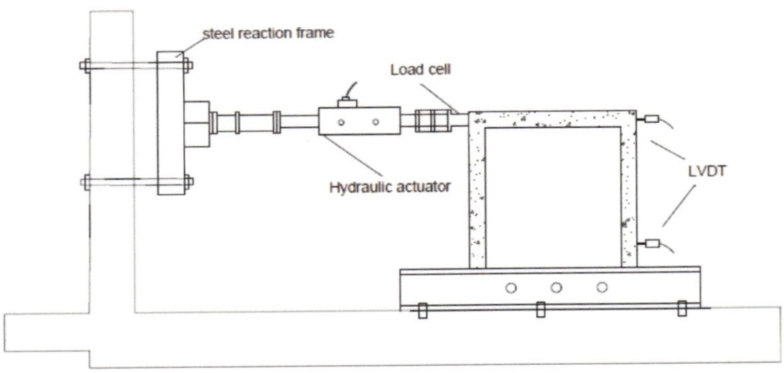

Figure 8.
Sketch of test setup.

3.5 Single bay frame without damper

Moreover, to estimate the efficiency of the MR damper, bare RCC frame without MR damper was also fabricated and tested as depicts in **Figures 9** and **10**. The loading pattern for the present experiment was depicted in **Figure 11**. From the results, it is clear that the first crack was observed in the 10th displacement cycle at the load of 6 kN applied loading pattern for the test experiment. The displacement was found to be 9.583 mm with the crack width of 1 mm. The maximum load carrying capacity was found to be 19 kN.

3.6 Frame with damper

To determine the ultimate load carrying capacity for the frame with MR damper. The experiment was carried out by supplying 3 A current to the MR damper placed in the RCC frame depicts in **Figures 12** and **13**. In the experiment, the first crack was observed at 16th displacement cycles at a load of 8kN which is higher than the bare frame. The displacement of 5.463 mm and a crack width was found to be 1 mm. The load carrying capacity was increased by 55% was observed for the frame with MR damper than the bare frame at 37.7 kN.

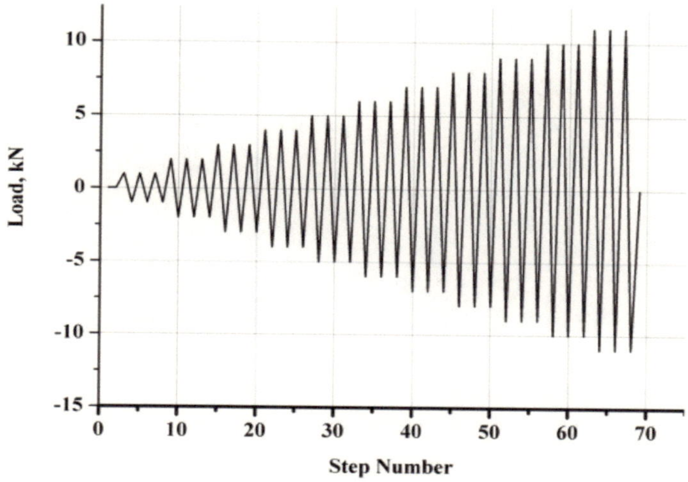

Figure 9.
Photograph of bare frame without damper.

Figure 10.
Load versus deflection for bare frame without MR damper.

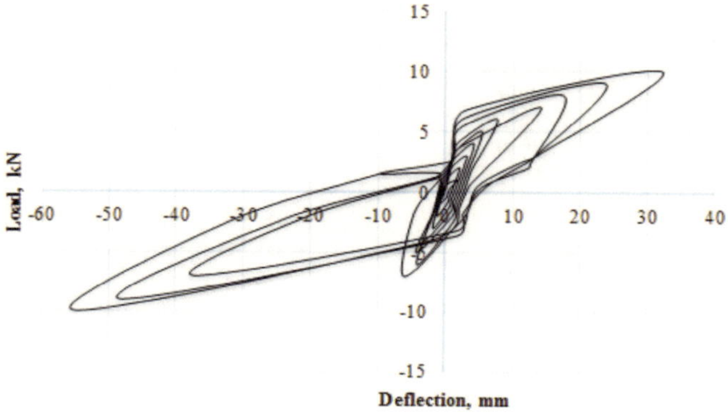

Figure 11.
Loading pattern of cyclic load test.

Figure 12.
Experimental setup of single bay frame with damper.

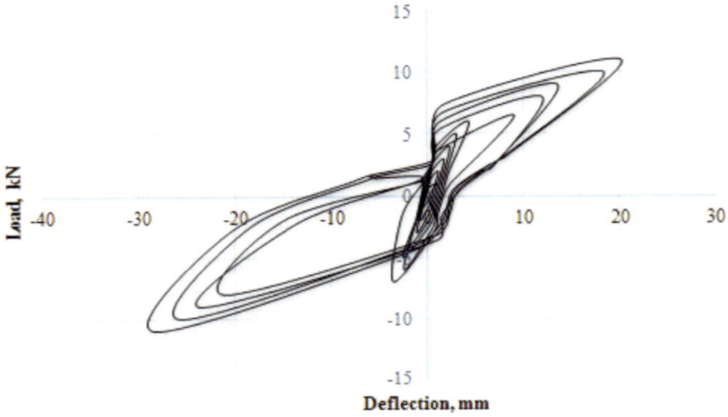

Figure 13.
Load versus deflection for frame with MR damper.

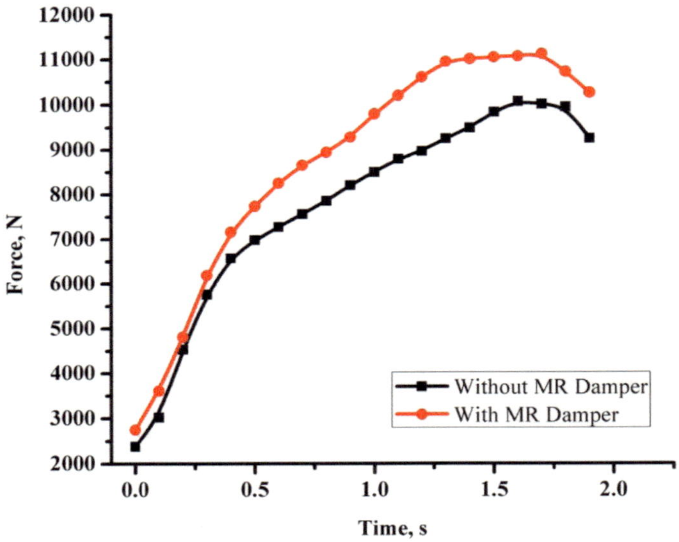

Figure 14.
Force versus time for with and without MR damper in RCC frame.

Figure 14 shows the force versus time curve for with and without MR damper. For this creep study the maximum load and deflection values for damper with and without RC frame are considered. The RCC frame with MR damper has the force of 10.24 kN. On comparing the force results for with and without MR damper, 9.79% of force was increased for RCC frame with MR damper.

Figure 15 shows the displacement versus time curve for with and without MR damper. The RCC frame with MR damper has the displacement of 19 mm. On comparing the displacement results for with and without MR damper, 39.85% of displacement was reduced for RCC frame with MR damper.

Figure 16 shows the stress versus time curve for with and without MR damper. The RCC frame with MR damper has the stress of 5124 N/mm^2. On comparing the

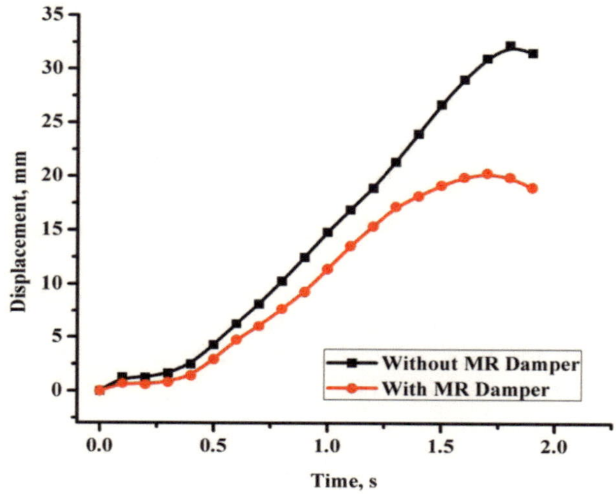

Figure 15.
Displacement versus time for with and without MR damper in RCC frame.

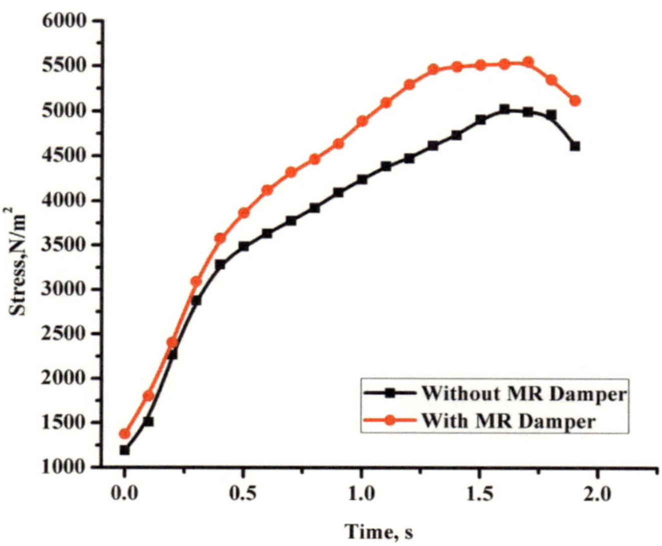

Figure 16.
Stress versus time for with and without MR damper in RCC frame.

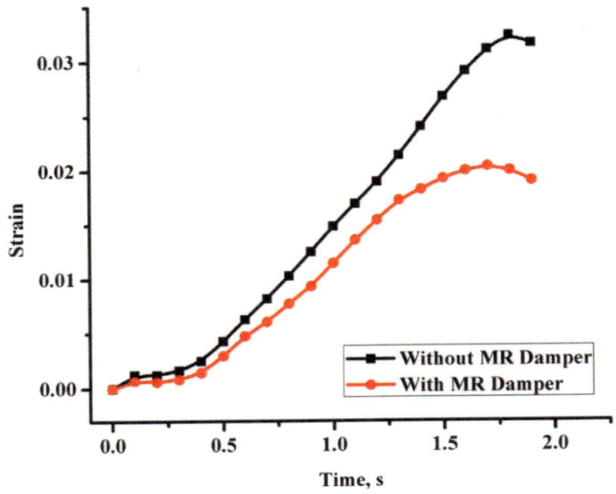

Figure 17.
Strain versus time for with and without MR damper in RCC frame.

strain results for with and without MR damper, 9.79% of strain was increased for RCC frame with MR damper.

Figure 17 shows the strain versus time curve for with and without MR damper. The RCC frame with MR damper has the stress of 0.019. On comparing the strain results for with and without MR damper, 38.09% of strain was decreased for RCC frame with MR damper.

4. Conclusions

Finally, it is concluded, for seismic resistance, MR damper are advantageous comparing to other supplementary dampers. In the present work, the developed MR damper with the MR fluid containing magnetic nano Fe_3O_4 and silicone oil as the carrier liquid was studied for MR damper application. On state and off state rheology of MR fluid was studied and its damping ability was compared. However, experimental investigation shows 76% of damping force was increased for the MR damper under on-state condition. From the cyclic load test, it is proven that 49% of displacement was reduced. The smart material, nano Fe_3O_4 was characterized using XRD and SEM and the average size of nano Fe_3O_4 was calculated and found to be 12 nm respectively. The experiment results reveal, the first crack was at 6 kN for RCC frame without MR damper and at 8 kN for the frame with MR damper. The ultimate value of the specimen reaches at 17 kN for frame without damper, MR damper it reaches a maximum load carrying capacity of 37.7 kN. The ultimate load carrying capacity and energy dissipation for RC frame with damper was increased by 55 and 45% respectively on compared to the bare frame. The effects of stress, strain, displacement and load behavior with respect to time for RC element with various time interval is observed with and without MR damper.

Acknowledgements

The authors thank the Karunya Institute of Technology and Sciences, Coimbatore, Tamil Nadu, India for their constant support. We also extend our acknowledgement to the Department of Science and Technology (Grant No: DST/TSG/STS/2015/30-G).

DOI: http://dx.doi.org/10.5772/intechopen.81971

Author details

Daniel Cruze[1*], Hemalatha Gladston[1], Sarala Loganathan[1], Tensing Dharmaraj[1] and Sundar Manoharan Solomon[2]

1 Karunya Institute of Technology and Sciences, Coimbatore, India

2 Sathyabama Institute of Science and Technology, Chennai, India

*Address all correspondence to: danielckarunya@gmail.com

References

[1] Hajalilou A, Mazlan SA, Lavvafi H, Shameli K. Magnetorheological (MR) fluids. In: Field Responsive Fluids as Smart Materials. Singapore: Springer; 2016. pp. 13-50. DOI: 10.1007/978-981-10-2495-5_3

[2] Zhang C, Wang LX. Dynamic models for magnetorheological dampers and its feedback linearization. Advanced Materials Research. 2009;**79**:1205-1208. DOI: 10.4028/www.scientific.net/amr.79-82.1205

[3] Mantripragada S, Wang X, Gordaninejad F, Hu B, Fuchs A. Rheological properties of novel magnetorheological fluids. International Journal of Modern Physics B. 2007;**21**(28-29):4849-4857. DOI: 10.1142/s021797920704575x

[4] Fujita T, Kabeya K, Hayamizu Y, Aizawa S, Higashino M, Kubo T, et al. Semi-active seismic isolation system using controllable friction damper: 1st report, development of controllable friction damper and fundamental study of semi-active control system. Transactions of the Japan Society of Mechanical Engineers. Series C. 1991;**57**(536):1122-1128. DOI: 10.1299/kikaic.72.1428

[5] Olabi AG, Grunwald A. Design and application of magneto-rheological fluid. Materials & Design. 2007;**28**(10): 2658-2664. DOI: 10.1016/j.matdes.2006.10.009

[6] Zapateiro M, Luo N, Taylor E, Dyke SJ. Experimental identification of a shear-mode MR damper and numerical evaluation of hysteretic models. In: ASME 2007 International Design Engineering Technical Conferences and Computers and Information in Engineering Conference; American Society of Mechanical Engineers. 2007. pp. 1469-1476. DOI: 10.1115/detc2007-35904

[7] Kim HS. Seismic response reduction of a building using top-story isolation system with MR damper. Contemporary Engineering Sciences. 2014;**21**:979-986. DOI: 10.12988/ces.2014.49123

[8] Li ZY, Dang PP, Mu DJ. MR damper semi active control for bridge under multi-support seismic excitation. Applied Mechanics and Materials. 2011;**90**:1402-1405. DOI: 10.4028/www.scientific.net/amm.90-93.1402

[9] Dey R, Saha P. Seismic response control of smart base-isolated benchmark building using hybrid control strategy (viscous fluid damper with MR damper). In: Recent Advances in Structural Engineering. Vol. 2. Singapore: Springer; 2019. pp. 365-374. DOI: 10.1007/978-981-13-0365-4_31

[10] Dyke SJ, Spencer BF Jr, Sain MK, Carlson JD. Seismic response reduction using magnetorheological dampers. IFAC Proceedings Volumes. 1996 Jun 1;**29**(1):5530-5535. DOI: 10.1016/s1474-6670(17)58562-6

[11] Dyke SJ, Spencer BF Jr, Sain MK, Carlson JD. Modeling and control of magnetorheological dampers for seismic response reduction. Smart Materials and Structures. 1996;**5**(5):565. DOI: 10.1088/0964-1726/5/5/006

[12] Kim Y, Langari R, Hurlebaus S. Seismic response control of a large civil structure equipped with magnetorheological dampers. In: Fuzzy Systems, 2009. FUZZ-IEEE 2009. IEEE International Conference on 2009 Aug 20; IEEE. pp. 215-220. DOI: 10.1109/fuzzy.2009.5277045

[13] Spencer BF, Dyke SJ, Sain MK. Magnetorheological dampers: A new approach to seismic protection of structures. In: Proceedings of the 35th IEEE Conference on Decision and

Control, 1996; IEEE. Vol. 1. pp. 676-681. DOI: 10.1109/cdc.1996.574412

[14] Yang G, Spencer BF Jr, Carlson JD, Sain MK. Large-scale MR fluid dampers: Modeling and dynamic performance considerations. Engineering Structures. 2002;**24**(3):309-323. DOI: 10.1016/s0141-0296(01)00097-9

[15] Hollenberg GW, Terwilliger GR, Gordon RS. Calculation of stresses and strains in four-point bending creep tests. Journal of the American Ceramic Society. 1971;**54**(4):196-199. DOI: 10.1111/j.1151-2916.1971.tb12263.x

[16] Piechnik S, Chrzanowski M. Time of total creep rupture of a beam under combined tension and bending. International Journal of Solids and Structures. 1970;**6**(4):453-477. DOI: 10.1016/0020-7683(70)90096-x

[17] Daniel C, Hemalatha G, Magdalene A, Tensing D, Manoharan SS. Magnetorheological damper for performance enhancement against seismic forces. In: International Congress and Exhibition Sustainable Civil Infrastructures-Innovative Infrastructure Geotechnology. Cham: Springer; 2017. pp. 104-117. DOI: 10.1007/978-3-319-61914-9_9

[18] Solomon SM, Dharmaraj T, Loganathan S, Immanuel S, Gladston H, Cruze D. Experimental investigation on magnetorheological damper for RCC frames subjected to cyclic loading. Advances in Civil Engineering Materials. 2018;**14**:7(3). DOI: 10.1520/acem20170112

[19] Seong M-S, Choi S-B, Sung K-G. Control Strategies for Vehicle Suspension System Featuring Magnetorheological (MR) Damper, Vibration Analysis and Control Francisco Beltran-Carbajal. IntechOpen; 6 Sep 2011. DOI: 10.5772/24556

[20] Rossi A, Orsini F, Scorza A, Botta F, Belfiore NP, Sciuto SA. A review on parametric dynamic models of magnetorheological dampers and their characterization methods. Actuators. 2018;**7**(2):16. DOI: 10.3390/act7020016

Chapter 3

Asphalt Material Creep Behavior

Amir Golalipour

Abstract

Asphalt binder, as one of the load-carrying components of the pavement, is a viscoelastic, thermoplastic material characterized by a certain level of rigidity of an elastic solid body, but, at the same time, flows and dissipates energy by frictional losses as a viscous fluid. Due to its complexity and importance, many studies were conducted to understand and alleviate its performance. Creep tests have been used to characterize asphalt materials at different service temperatures. Permanent strain or rutting is one of the most important pavement distresses. It is believed that the accumulated strain in asphalt binder, as a consequence of traffic, is mainly responsible for the rutting of asphalt pavements. Repeated creep tests were developed to identify non-viscous flow that contributes to the permanent deformation from the total dissipated energy. The low-temperature cracking of asphalt pavements is a major pavement distress mechanism in cold regions. Since asphalt is a viscoelastic material, part of said stresses is dissipated through relaxation, but, eventually, they build up until they reach the strength of the material, leading to the formation of cracks to relieve these stresses. Conducting creep test at low temperatures is a common test method to characterize thermal cracking behavior of asphalt binders.

Keywords: asphalt, pavements, creep test, permanent strain, rutting, thermal cracking

1. Asphalt rutting resistance

Asphalt pavements are granular composites that contain mineral aggregates, asphalt binder and air voids. The two load-carrying components of the asphalt mixtures are the mineral aggregates and the binder. Asphalt binders are obtained from the refining of crude oil. They are produced from the heavy residue after the refining of fuels and lubricants. Asphalt is a thermoplastic material that demonstrates viscoelastic properties under most pavement operating conditions [1]. It is this fundamental property that makes them versatile binders for asphalt mixtures with the viscoelastic characteristics of the bituminous binders directly and significantly influencing the performance of the mixtures.

One of the distress modes of asphalt pavements is permanent deformation or rutting that occurs at high operating temperatures, and it is believed that the accumulated strain in asphalt binder is mainly responsible for the rutting. Rutting is defined as longitudinal surface depressions along a pavement's wheel paths. In asphalt pavements, rutting is defined as the progressive accumulation of longitudinal depressions in a wheel path under repetitive loading [2].

The majority of rutting problems result from plastic deformation of the surface course. It is characterized by shear deformation inside the asphalt mixture [3]. The permanent deformation per wheel passage correlates with the stiffness of the asphalt binder used, and decreases with increasing number of wheel passages [4]. Also, permanent deformation in asphalt binder is highly dependent on the stress level. One of the most recent areas of investigation in pavement engineering is the relationship between asphalt rheology and rutting in pavements.

Asphalt mixture is a viscoelastic material that contains mineral aggregates, asphalt binder and air voids. Asphalt binder, as one of the load carrying components of the asphalt mixtures, is a viscoelastic, thermoplastic material that is characterized by a certain level of rigidity of an elastic solid body, but, at the same time, flows and dissipates energy by frictional losses as a viscous fluid [5]. Its characteristics are dependent on time and temperature [6]. At higher temperatures and longer loading times, the asphalt material softens and behaves more similar to a viscous fluid. At lower temperatures and faster loads, the asphalt material becomes stiffer and act more as an elastic material. Because of this, rutting is more critical during the hot season in a year and under slower moving traffic.

As the asphalt binder is responsible for the viscoelastic behavior of all bituminous materials, it plays a dominant part in determining many of the aspects of pavement performance, such as resistance to permanent deformation. Therefore, binder has a critical role against rutting in mixture. Also, as with any viscoelastic material, asphalt's response to stress is dependent on both temperature and loading time [5]. Therefore, permanent deformation in asphalt binder is highly dependent on the factors such as temperature, stress level, loading time etc.

1.1 Asphalt binder characterization and tests at high temperatures

Recognizing the limitation of the traditional asphalt binder characterization procedure in 1987, the Federal Highway Administration initiated a nationwide research program called the Strategic Highway Research Program, usually referred to as SHRP [6]. The final product of the SHRP research program was Superpave® (Superior Performance Asphalt Pavements). The Superpave® was designed to provide performance-related properties that can be related in a rational manner to pavement performance [5].

The Dynamic Shear Rheometer (DSR) was introduced in 1993 by Superpave® as a tool to measure the binder mechanical characterization. This device provided a useful method to evaluate binder rutting resistance capability. The principle used with the DSR is to apply sinusoidal, oscillatory stresses or strains over a range of temperatures and loading frequencies to a thin disc of bitumen, sandwiched between the two parallel plates of the DSR. Anderson et al. [6] assumed that rutting is caused by the total dissipated energy as calculated from the strain-stress curve.

$$W_i = \pi * {\tau_0}^2 * \frac{1}{\frac{G^*}{\sin\delta}} \quad (1)$$

where:

W_i = total energy dissipated at the ith cycle, τ_0 = maximum stress applied, G^* = complex modulus, δ = phase angle.

$|G^*|/\sin\delta$ was introduced as the rutting parameter. Equation 1 shows that increasing the rutting parameter $|G^*|/\sin\delta$ causes dissipated energy to decrease and, as a consequence, more rutting occurs.

Using cyclic reversible loading for viscoelastic materials can be misleading because although this test has the capability of estimation the total energy dissipated during a loading cycle, as it is unable to separate permanent deformation and delay elasticity in these materials. Rutting is a repeated mechanism with sinusoidal loading pulse in which the pavement layer is not forced back to zero deflection but would recover some deformation due to elastic stored energy in the material of the layers. Under this type of loading, the energy is dissipated in damping and in permanent flow [7]. It was proposed that a creep and recovery test in the DSR could solve the above problems [8].

1.2 Repeated creep and recovery (RCR) test

The repeated creep test is proposed as a method of separating the dissipated energy and estimating the resistance to accumulation of permanent strain for asphalt binders. During the NCHRP 9-10 project, Bahia et al. [8] suggested the repeated creep recovery test (RCR) as a better tool to investigate the rutting resistance of asphalt binders using the dynamic shear rheometer (DSR). The NCHRP 9-10 project recommend a shear stress in the range of 30–300 Pa for 100 cycles at a rate of 1 (second) loading time followed by a 9 (seconds) unloading time [8].

This project introduced a new parameter G_v to characterize the rutting resistance of asphalt binders. This new parameter was derived from the four-element Burger model, which is a combination of a Kelvin model and Maxwell model. The total shear strain versus time is expressed as follows:

$$\gamma\,(t) = \gamma_1 + \gamma_2 + \gamma_3 = \frac{\tau_0}{G_0} + \frac{\tau_0}{G_1}\left(1 - e\frac{-tG_1}{\eta_1}\right) + \frac{\tau_0}{\eta_0}t \tag{2}$$

where γ (t) = total shear strain, γ_1 = elastic shear strain, γ_2 = delayed elastic strain, γ_3 = viscous shear strain, τ_0 = constant stress, G_0 = spring constant of Maxwell model, tG_1 = spring constant of Kelvin model, η_1 = dashpot constant of Kelvin model, t = time, η_0 = dashpot constant of Maxwell model.

Dividing Eq. (2) by the constant stress leads to the following equation:

$$J(t) = J_e + J_{de}(t) + J_v(t) \tag{3}$$

where J_e = elastic creep compliance, $J_{de}(t)$ = delayed elastic creep compliance, $J_v(t)$ = viscous creep compliance.

Instead of using J_v which has a unit of 1/Pa, the inverse of compliance G_v is used. G_v is defined as the viscous component of the creep stiffness [8].

D'Angelo showed that a single stress level did not completely account for the stress dependency of modified binders and multiple stress levels need to be used [9]. Testing binders at multiple stress level using the RCR test would require an extensive amount of time.

NCHRP 9-10 project recommended RCR testing using the DSR to evaluate the rutting resistance of asphalt binders using G_v parameter showing viscous component of the creep stiffness. This parameter has shown promising results to characterize the rutting behavior of asphalt binder, especially modified binders. The main benefits of RCR compared with the current test standard and parameter $|G^*|/\sin\delta$ are described in previous work [10] and can be summarized as follows:

- Better simulation of actual loading from truck on the pavement using repeated creep loading compared to dynamic shear procedure.

- It is simpler way to identify binder rutting resistance using permanent deformation derived from RCR test. This method highlights the effect of delayed elasticity which is very important for modified binders and allows the binder to recover deformation during the rest period. The parameter $|G^*|/\sin\delta$ is unable to directly evaluate the delayed elasticity.

It is not fully understood if testing at low stress levels is representative of binder rut resistance behavior as happens in the field since the stresses and strains in the binder can be very high and non-linear. Permanent deformation in asphalt binder is highly dependent on the stress level. Determining the stress level at which the binder is exposed in the mixture is an important matter [11]. Permanent deformation is not a linear viscoelastic phenomenon and therefore measurement of linear viscoelastic binder properties are not likely to correlate with it [12].

The RCR test provides valuable information about the asphalt material rutting resistance when temperature changes happen. However, since the loading on pavement in the field is not consistent, multiple stresses are required to accurately characterize the asphalt binder rutting behavior. These stresses should be selected to capture the properties of the asphalt in linear and in nonlinear domain.

1.3 Multiple stress creep and recovery (MSCR)

The MSCR test was developed to reduce the number of samples at each stress level and it is the following development of RCR test. The test uses 1 (second) creep loading followed by 9 (seconds) recovery for the following stress levels: 25, 50, 100, 200, 400, 800, 1600, 3200, 6400, 12,800 and 25,600 Pa at 10 cycles for each stress level. The test starts at the lowest stress level and increase to the next stress level at the end of every 10 cycles, with no rest periods between creep and recovery cycles or changes in stress level [11].

D'Angelo selected two stress levels, 0.1 and 3.2 kPa, upon correlation between binder and mixture rutting results for performing the MSCR test. Ten cycles are run for each stress level for a total of 20 cycles. **Figure 1** shows the typical results from MSCR test.

The average non-recoverable strain for the 10 creep and recovery cycles is then divided by the applied stress for those cycles yielding the non-recoverable creep compliance (J_{nr}). J_{nr} for 0.1 kPa is calculated by divided the strain after 10 cycles to 0.1 kPa. Equations (4) and (5) show the calculation method for Jnr at 0.1 kPa.

$$J_{nr}(0.1, N) = \frac{\varepsilon_{10}}{0.1} \tag{4}$$

$$J_{nr0.1} = \frac{SUM[J_{nr}(0.1, N)]}{10} \; for\; N = 1 \; to \; 10 \tag{5}$$

where $\varepsilon_{10} = \varepsilon r - \varepsilon o$, εr = strain value at the end of the recovery portion (i.e., after 10.0 seconds) of each cycle strain, εo = initial strain value at the beginning of the creep portion of each cycle.

The definition for the J_{nr} at 3.2 kPa is analogous. The J_{nr} parameter was suggested as a measure of the binder contribution to mixture permanent deformation [13].

Later, Shenoy [14] used non-recoverable compliance to characterize the propensity of asphalt binder to resist permanent deformation in the pavement. He proposed measuring the non-recoverable compliance through the dynamic oscillatory test using a frequency, time, strain or sweep test. Mathematical formula was used to calculate non-recoverable compliance from the complex modulus and phase angle. Also Shenoy showed that the unrecoverable strain in a binder that is during a creep and recovery test could be calculated directly from the dynamic oscillatory

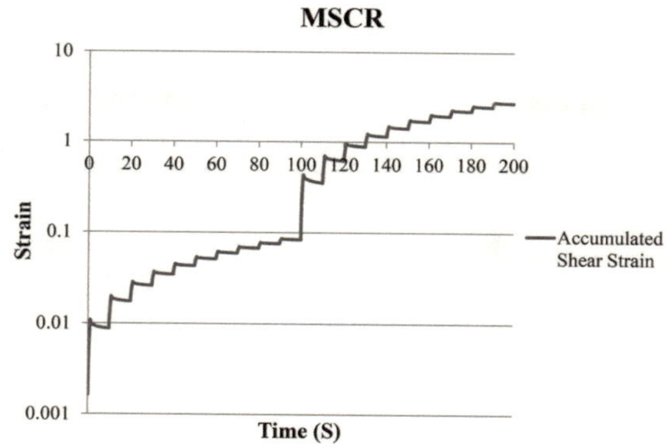

Figure 1.
MSCR response curve.

test using a frequency test [15]. The percent-unrecovered strain is calculated as follows:

$$\%\gamma_{unr} = \frac{100\ \sigma_0}{|G^*|}\left(1 - \frac{1}{tan\delta sin\delta}\right) \tag{6}$$

To minimize the unrecoverable strain, the following term (the inverse of the non-recoverable compliance, $\%\gamma_{unr}/\sigma_o$) needs to be maximized:

$$\frac{|G^*|}{1 - \frac{1}{tan\delta sin\delta}} \tag{7}$$

|G*|/ (1−(1−tanδsinδ)) was proposed as a refinement to the Superpave® specification for performance grading of asphalts. The drawback of this parameter is that at $\delta < 52°$, the model would predict unrealistic negatives values of (1−(1−tanδsinδ)) [15].

Bouldin et al. [16] developed a semi empirical method to predict the rut resistance (R) as a function of loading (time and load) and temperature from data at single frequency. This approach is based on the assumption that the strain accumulation rate depends on the binder stiffness and viscoelastic contribution f(δ), and these two contributions are independent

$$R = (1/\gamma_{acc})(t_{load}, t_{rest}) = G*(\omega)f[\delta(\omega)] \tag{8}$$

$$(1/\gamma_{acc}) = k\ G*\left(Y_0 + a\left\{1 - \exp\left[-|\delta - X_0 + b\ln(2)^{1/c}|/b\right]^c\right\}\right) \tag{9}$$

where k = constant, γ_{acc} = accumulated strain, δ = phase angle, Y_0, X_0, a, b, c = empirical fitting parameters.

The disadvantage of this parameter is that at phase angles between 40 and 75° this parameter does not fully capture the viscoelastic nature of many modified binders [16].

The MSCR is proposed as a better test to evaluate modified binders and estimate their role in pavement performance. The MSCR has received many positive feedbacks as a good candidate for rutting evaluation and shown a great promise. However, this test has faced few challenges regarding implementation, especially in terms of analysis of the result and interpretation.

While results of the MSCR test are promising [9, 11, 13, 17, 18], there are important concerns about current testing and analysis protocols. It is not clear that testing at a low stress levels (0.1 and 3.2 kPa) is the best way to characterize the rutting resistance of an asphalt binder. The stresses and strains in the binder can be high, much higher than the linear limit for the material. Permanent deformation in asphalt binder is highly dependent on the stress level. Determining the stress level at which the binder is exposed in the mixture is an important matter [19, 20]. Permanent deformation is not a linear viscoelastic phenomenon and, therefore, measurement of linear viscoelastic binder properties are not likely to correlate with it [21]. The selection of two stress levels is not necessarily based on the stresses that asphalt binder experiences inside the pavement. The number of cycles and the time of loading do not provide full picture of characterizing long term deformation in the material. The recovery time may need to be longer to fully capture delayed elastic behavior of modified binders; some binders are still recovering after 9 (seconds) of recovery [22]. Golalipour [23] investigated these factors and provided some improvements for the MSCR test protocol.

In the latest version of AASHTO standard T 350 standard "Standard Method of Test for Multiple Stress Creep Recovery (MSCR) Test of Asphalt Binder Using a Dynamic Shear Rheometer (DSR)" or American Society for Testing and Materials standard (ASTM) 7405, the test consists of 20 cycles of 0.1 kPa stress creep and recovery, followed immediately by another 10 cycles of 3.2 kPa stress creep and recovery. Each cycle consists of 1 second of loading and 9 seconds of recovery upon instantaneous unloading [24, 25].

The non-recoverable creep compliance, J_{nr}, and the percent recovery, R%, are two of the parameters calculated from the measured strain under different stress cycles [9, 11]. The J_{nr} parameter was suggested as a measure of the binder contribution to mixture permanent deformation. Different factors can have significant impact on J_{nr} value such as the duration of the creep interval, the duration of the recovery interval, the number of loading cycles and, of course, the entity of the applied shear stress. In other words, J_{nr} depends on the mechanical history of the experiment.

2. Asphalt thermal cracking behavior

The low-temperature cracking of asphalt concrete pavements is a major pavement distress mechanism in cold regions costing hundreds of millions of dollars in rehabilitation costs to various agencies. It usually occurs in the form of regularly spaced transverse cracks, initiating at the surface of the asphalt layer and further propagating downward. Consequences of thermal cracking are an immediate increase of the roughness of the pavement surface (i.e., a reduction of the comfort and safety of the ride) and the loss of the sealing function of the pavement for the underlying layers. However, predictions of this distress have not been accurate enough, often resulting in premature road failures. It is believed that the excessive brittleness due to the increase in stiffness and decrease in the ability to relax stress leads to the buildup of thermally induced stress and ultimately cracking of mixtures in pavements.

The prediction of asphalt pavement thermal cracking has been the subject of numerous studies that date back to the early 1960's [26]. In many of these studies, attempts were made to introduce a procedure to predict pavement cracking based on the stress-strain-time—temperature relationship. A number of methods have been introduced throughout the years to model the viscoelastic behavior of asphalt binders and mixtures to estimate the accumulation of thermal stress during cooling cycles and predict the temperature at which cracking occurs [27–29].

2.1 Asphalt binder characterization and tests at low temperatures

It has been shown that the characteristic of the asphalt binder can significantly impact the low temperature behavior of asphalt pavement. During the service life of the pavement, asphalt is exposed to low temperatures, which tend to alter the rheological behavior. Different studies have concluded that asphalt binder behavior is the dominant component for low-temperature performance of the asphalt pavement mixtures [30]. Therefore, it is very important to study the asphalt binder characterize at low temperatures to have a clear picture of factors that affect the low temperature behavior of pavement.

Based on the concept that asphalt binder properties play the major role in cracking, several studies have focused on investigating the effect of rheological and other parameters of asphalt binder on low temperature performance. A quantitative method is necessary in order to study the complex role of asphalt binder in the pavement and to relate its properties to low temperature cracking phenomenon.

It is believed that thermal stresses causing cracking are due to constrained thermal strains. When the temperature drops, the pavement tends to contract its volume, following its thermal expansion/contraction coefficient. However, the layer underneath opposes some resistance due to friction, therefore thermal strains within the asphalt layer are not free to take place leading to co-active stresses proportional to the stiffness of the material. Since asphalt is a viscoelastic material, part of said stresses are dissipated through relaxation, but eventually they build up until they reach the strength of the material, leading to the formation of cracks to relieve these stresses [31].

A number of methods have been introduced throughout the years to model the viscoelastic behavior of asphalt binders and mixtures to estimate the accumulation of thermal stress during cooling cycles and predict the temperature at which cracking occurs. Shoor et al. [32] studied the penetration measurement to investigate the low temperatures behavior of asphalt binder. Majidzadeh and Schweyer [33] investigated asphalt low temperature properties using more fundamental approach using viscoelastic models. They studied few asphalt binders' behavior at the temperature range of −9–5°C using cylindrical specimens. At the lowest temperatures, the asphalt binders exhibited some instantaneous elastic deformation. Pink et al. [34] used a Rheometric Mechanical Spectrometer (RMS) to make accurate low-temperature viscoelastic measurements on asphalt down to −94°C. They developed a methodology to construct a dynamic master curve to separate the effect of time and temperature. Button et al., also used the RMS to measure viscosity of asphalts from 0 to −46°C. Thus, most of the predicting low-temperature cracking methods involve measured asphalt stiffness, predicted asphalt stiffness, or consistency and temperature susceptibility parameters that indirectly establish asphalt stiffness [35].

During the SHRP project, several studies have focused on developing a new device for measuring low temperature stiffness of binders. Finally, these studies led to the development a device to determine the properties and response of asphalt binders at low temperatures in the 1980's. This device was later modified and updated as part of the SHRP binder research [36]. The resulting machine was named the Bending Beam Rheometer (BBR).

2.2 Bending beam rheometer (BBR) test

The data acquisition system of the BBR records the load and deflection test results and calculates two parameters: (1) Creep Stiffness, S(t), which is a measure of how the asphalt binder resists the creep loading, and (2) m-value, which is a measure of the rate at which the creep stiffness changes with loading time (**Figure 2**).

Thermo-mechanical properties of the asphalt samples can be measured using the Bending Beam Rheometer to evaluate the low temperature properties based on the ASTM D6648 and AASHTO T 313 [37, 38]. The stiffness, S(t), is a measure of the thermal stresses developed in the asphalt pavements as result of thermal contraction. Classic beam analysis theory is used to calculate the creep stiffness of the asphalt binder beam at 60 seconds loading time [5]. The BBR loads the beams for 240 seconds and report the stiffness values at loading times of 8, 15, 30, 60, 120 and 240 seconds. These values were chosen because they are fairly equally spaced on a logarithmic time scale. These data points, along with the following equation, are used to determine the shape of the stiffness (creep compliance) master curve for the asphalt binder (Eq. (10)).

$$S(t) = A + B\ \log(t) + C\,[\log(t)]^2 \tag{10}$$

where S(t) = asphalt binder stiffness, T = time (s), A, B and C = constants.

The slope of the stiffness curve, m, is a measure of the rate of stress relaxation by asphalt binder flow. The m-value indicates the rate of change of the stiffness, S(t), with loading time. In other words, the m-value is the slope of the log creep stiffness versus log time curve at any time. Since the time dependency of asphalt binder varies, the shape of the stiffness master curve as well as the stiffness at 2 hours loading time are important to take into consideration. Therefore, the slope of the stiffness master curve is also used for specification purposes [39].

In the current PG specification, two parameters from BBR test are used to characterize the binder low temperature rheological behavior. Apparent stiffness, S, and the slope of the log stiffness versus log time, the m-value are determined at a loading time of 60 seconds. The temperature at which $S(60) \leq 300$ MPa and m $(60) \geq 0.3$ is specified as the critical temperature + 10°C (**Figure 3**). These limits were established based on data from previous studies as well as the data obtained by SHRP (Bahia and [6]).

The effect of these two specification parameters, S(t) and m-value, on thermal cracking is analogous to the effect of G* and δ on rutting and fatigue cracking. As S(t) increases, the thermal stresses developed in the pavement due to thermal shrinking also increase, and thermal cracking becomes more likely. On the other hand, as the m-value decreases, so does the rate of stress relaxation. In other words,

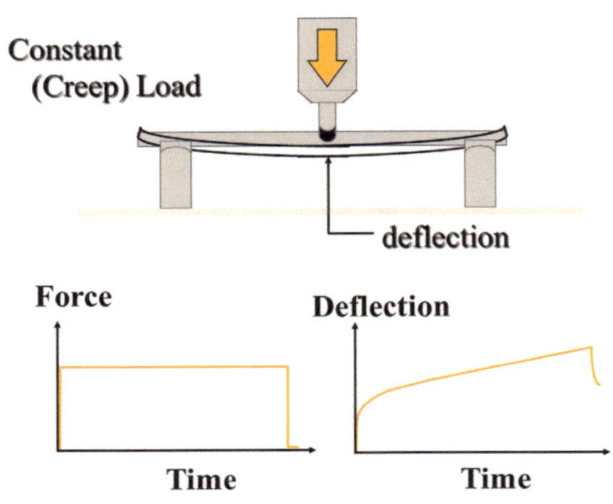

Figure 2.
BBR test principles.

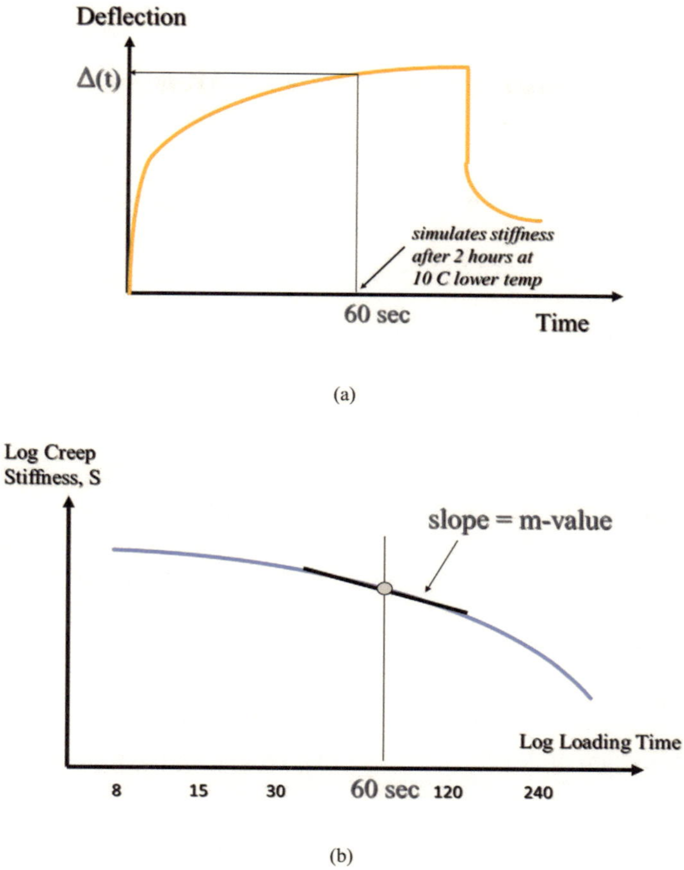

Figure 3.
BBR data analysis: (a) stiffness; (b) m-value.

as the slope of the asphalt binder stiffness curve flattens, the ability of the asphalt pavement to relive thermal stresses by flow decreases. This again would increase the propensity of thermal cracking in the pavement.

Author details

Amir Golalipour
Federal Highway Administration (FHWA)/Engineering and Software Consultants, Inc., McLean, VA

*Address all correspondence to: amir.golalipour.ctr@dot.gov

References

[1] Van der Poel C. A general system describing the visco-elastic properties of bitumen and its relation to routine test data. Journal of Applied Chemistry. 1954;**4**:221-236

[2] Faheem A, Bahia H. Using gyratory compactor to measure mechanical stability of asphalt mixtures. Wisconsin Highway Research Program; 2004

[3] Centeno M, Sandoval I, Cremades I, Alarcon J. Assessing rutting susceptibility of five different modified asphalts in bituminous mixtures using rheology and wheel tracking test (No. 08-0705); 2008

[4] Hofstra A, Klomp AJG. Permanent deformation of flexible pavements under simulated road traffic conditions. In: Third International Conference on the Structural Design of Asphalt Pavements. Vol. 1. 1972. pp. 613-621

[5] McGennis RB. Background of SUPERPAVE asphalt mixture design and analysis. Federal Highway Administration; Available through the National Technical Information Service; 1994

[6] Anderson DA, Peterson JC, Christensen DW. Variations in asphalt cements and their effects on performance of asphalt concrete mixtures. In: Presented to the Association of Asphalt Paving Technologies; 1986

[7] Kim YR. Modeling of Asphalt Concrete (the American Society of Civil Engineers). McGraw-Hill Professional; 2009

[8] Bahia HU, Hanson DI, Zeng M, Zhai H, Khatri MA and Anderson MR. NCHRP Report 459, Characterization of Modified Asphalt Binders in Superpave Mix Design. Prepared for the National Cooperative Highway Research Program. Washington, D.C.: Transportation Research Board, National Research Council; 2001

[9] D'Angelo JA, Dongre R, Reinke G. Creep and recovery. Public Roads. 2007; **70**(5):24-30

[10] Delgadillo R, Nam K, Bahia HU. Why do we need to change G*/sin δ and how? Road Materials and Pavement Design. 2006;**7**(1):7-27

[11] Delgadillo R, Cho DW, Bahia HU. Nonlinearity of repeated creep and recovery binder test and relationship with mixture permanent deformation. Transportation Research Record. 2006b; **1962**(1):2-11

[12] D'Angelo J, Kluttz R, Dongre R, Stephens K, Zanzotto L. Revision of the superpave high temperature binder specification: The multiple stress creep recovery test. Asphalt Paving Technology. 2007b;**76**(123)

[13] D'Angelo R, Dongre R, Reinke G. Evaluation of repeated creep and recovery test method as an alternative to SHRP+ requirements for polymer modified asphalt binders. In: Proceedings, Canadian Technical Asphalt Association, November; 2006

[14] Shenoy A. Refinement of the superpave specification parameter for performance grading of asphalt. Journal of Transportation Engineering. 2001;**127** (5):357-362

[15] Shenoy A. Nonrecovered compliance from dynamic oscillatory test Vis-a-Vis nonrecovered compliance from multiple stress creep recovery test in the dynamic shear rheometer. International Journal of Pavement Engineering. 2008;**9**:329-341

[16] Bouldin MG, Dongré R, D'Angelo J. Proposed refinement of superpave high-

temperature specification parameter for performance-graded binders. Transportation Research Record: Journal of the Transportation Research Board. 2001;**1766**(1):40-47

[17] D'Angelo R, Dongre R. Practical use of the MSCR test: Characterization of SBS dispersion and other additives in PMA binders. In: Transportation Research Board Annual Meeting; 2009

[18] D'Angelo R, Dongre R. Development of a high temperature performance based binder specification in the United States. In: 10th International Conference on Asphalt Pavements, Quebec City, Canada, 10th International Conference on Asphalt Pavements; 2006b

[19] Dreessen A, Planche JP, Gandel V. A new performance related test method for rutting prediction: MSCRT. In: Loizos A, Partl MN, Scarpas T, Al-Qadi IL, editors. Advanced Testing and Characterization of Bituminous Materials. Vol. 2. London: CRC Press, Taylor & Francis Group; 2009

[20] Wasage TLJ, Stastna J, Zanzotto L. Rheological analysis of multiple stress creep and recovery (MSCR) test. International Journal of Pavement Engineering. 2011

[21] William ML, Landel RF, Ferry JD. The temperature dependence of relaxation mechanisms in amorphous polymers and other glass-forming liquids. Journal of the American Chemical Society. 1955;77:3701-3706

[22] Delgadillo R. Nonlinearity of asphalt binders and the relationship with asphalt mixture permanent deformation [Ph.D Thesis]. Madison: University of Wisconsin-Madison; 2008

[23] Golalipour A. Modification of multiple stress creep and recovery test procedure and usage in specification [Master Thesis]. University of Wisconsin; 2011

[24] AASHTO T 350. Standard method of test for multiple stress creep recovery (MSCR) test of asphalt binder using a dynamic shear rheometer (DSR). Washington, D.C: American Association of State Highway and Transportation Officials; 2015

[25] ASTM D7405. Standard Method of Test for Multiple Stress Creep and Recovery (MSCR) of Asphalt Binder Using a Dynamic Shear Rheometer. West Conshohocken, Pennsylvania: American Society for Testing and Materials; 2016

[26] Monismith CL, Secor GA, Secor KR. Temperature-induced stresses and deformation in asphalt concrete. In: Proceedings of the Association of Asphalt Paving Technologists. Vol. 34. 1965. pp. 248-285

[27] Haas R. A Method for Designing Asphalt Pavements to Minimize Low-Temperature Shrinkage Cracking. RR-73-1, Asphalt Institute; 1973

[28] Lytton RL, Uzan J, Fernando EG, Roque R, Hiltunen D, Stoffels SM. Development and validation of performance prediction models and specifications for asphalt binders and paving mixes. SHRP-A-357, Strategic Highway Research Program, National Research Council; Washington, D.C; 1993

[29] Ruth BE, Bloy LL, Avital AA. Prediction of pavement cracking at low temperatures. In: Asphalt Paving Technology, Asphalt Paving Technology, Proceedings, the Association of Asphalt Paving Technologists. Vol. 51. 1982. pp. 53-103

[30] Golalipour A, Bahia H. Evaluation of oil modification effect on asphalt binder thermal cracking and aging properties. In: Proceedings of the Fifty-

Ninth Annual Conference of the Canadian Technical Asphalt Association (CTAA): Winnipeg, Manitoba; 2014

[31] Ruth BE. Prediction of low-temperature creep and thermal strain in asphalt concrete pavements. In: Mack CR, editor. American Society for Testing Materials; 1977. pp. 68-63. ASTM STP 628

[32] Shoor K, Majidzadeh K, Schweyer HE. Temperature-flow functions for certain asphalt cements. In: Highway Research Record 134, HRB, National Research Council, Washington, D.C.; 1966. pp. 63-74

[33] Majidzadeh K, Schweyer HE. Free volume considerations for asphalts. Materials Research and Standards. 1966; **6**:617

[34] Pink HS, Merz RE, Bosniak DS. Asphalt rheology: experimental determination of dynamic Moduli at low temperature. In: Proceedings of the Association of Asphalt Paving Technologies. Vol. 49. 1980. pp. 64

[35] Anderson DA, Christensen DW, Bahia HU, Dongre R, Sharma MG, Antle CE, et al. Binder Characterization and Evaluation. Vol. 3: Physical Characterization. Strategic Highway Research Program, National Research Council, Washington, DC; 1994

[36] Bahia HU. Low-temperature isothermal physical hardening of asphalt cements. Ph.D. dissertation. Pennsylvania State University; 1991

[37] AASHTO T 313. Standard method of test for determining the flexural creep stiffness of asphalt binder using the bending beam rheometer (BBR). Washington, D.C: American Association of State Highway and Transportation Officials; 2010

[38] ASTM D6648. Standard Test Method for Determining the Flexural Creep Stiffness of Asphalt Binder Using the Bending Beam Rheometer (BBR). West Conshohocken, Pennsylvania: American Society for Testing and Materials; 2008

[39] Anderson DA et al. Binder characterization and evaluation. Vol. III. Final Report for SHRP Project A-002A; 1993

Chapter 4

Creep Characteristics of Engineering Materials

Janusz Dobrzański and Adam Zieliński

Abstract

The assessment of suitability for further service of the basic material and circumferential welded joints of primary steam pipelines made of 14MoV6-3 steel was made based on the determined characteristics of their mechanical properties and structure. The tests were carried out on materials after service significantly exceeding the design time and 200,000 h. The mechanical properties with respect to the determined class of structure for the basic and welded joint material were compared to the determined class of damage in accordance with the Institute for Ferrous Metallurgy's own classification of the condition of materials after service under creep conditions. The residual life and disposable residual life were determined based on short-term creep tests and referred to the class of existing structure and the class of damage. The assessment of the suitability of the basic and circumferential welded joint material for further service was made.

Keywords: creep, welded joint, Cr-Mo-V steel, microstructure, residual life

1. Introduction

In the European conventional power generation, a small number of newly constructed units have been recorded in at least several recent years, which is caused by, but not limited to, a lot of factors related mainly to environmental protection. Therefore, the electricity producers direct their main efforts at maintaining the availability of existing power units while ensuring their safe service. The inspections of and repairs to the energy installations that are in use and whose design service life was in most cases significantly exceeded while at the same time providing the reliable efficient diagnostics seem to be the proper way for the achievement of these objectives.

However, the performance of periodic inspections and repairs is not sufficient to maintain the current level of electricity and heat production in Poland. The modernisation of the units being operated is required as well. To maintain the current level of production, it is estimated that each year the modernisation of units with a total capacity of at least 2000 MW is necessary. Therefore, the important issue to be resolved has become not only the condition assessment of the pressure section of boiler and forecast about its further safe service but also the design modernisation with selection and use of new materials allowing for construction and installation of new components [1–10]. The required increase in efficiency and extension of service time of the operating high-power units, i.e. 200, 360 and 500 MW, far beyond the design time up to the expected min. 350,000 h involves the need to replace certain components because of the end of their life and to modernise them. Certain critical components being replaced are often made of new-generation steel for the power generation industry [11–20]. At the same time, these units are expected to be operated in the control system and to comply with standards of the European directive on limitation of harmful emissions into the atmosphere.

As part of the diagnostics, it is necessary to perform the condition assessment not only for the basic material of the components working under creep conditions but also for the material of welded joints. The condition assessment of the welded joint material is necessary to determine the ability of the components to transfer the required service loads during their further service [21–34].

If there is a need to repair or replace a part or entire component with a new one, the capability of the basic material of the operated component to make such a repair or replacement should be determined. When condition of this material after service allows a repair, it is necessary to develop the repair technology from time to time.

The subject matter of the testing, including non-destructive and destructive tests, related to the long-time operated and repair welded joints of materials after long-term service with materials after service or with new materials is a separate issue dealt with by the Institute for Ferrous Metallurgy for many years.

The developed and applied research methods and the adopted procedure allow for the correct condition assessment of the basic material and welded joint material (HAZ and weld). The test results for long-time operated as well as repair and modernisation welded joints obtained so far allowed expanding the material database with such data and proposing the procedure for making such welded joints when necessary.

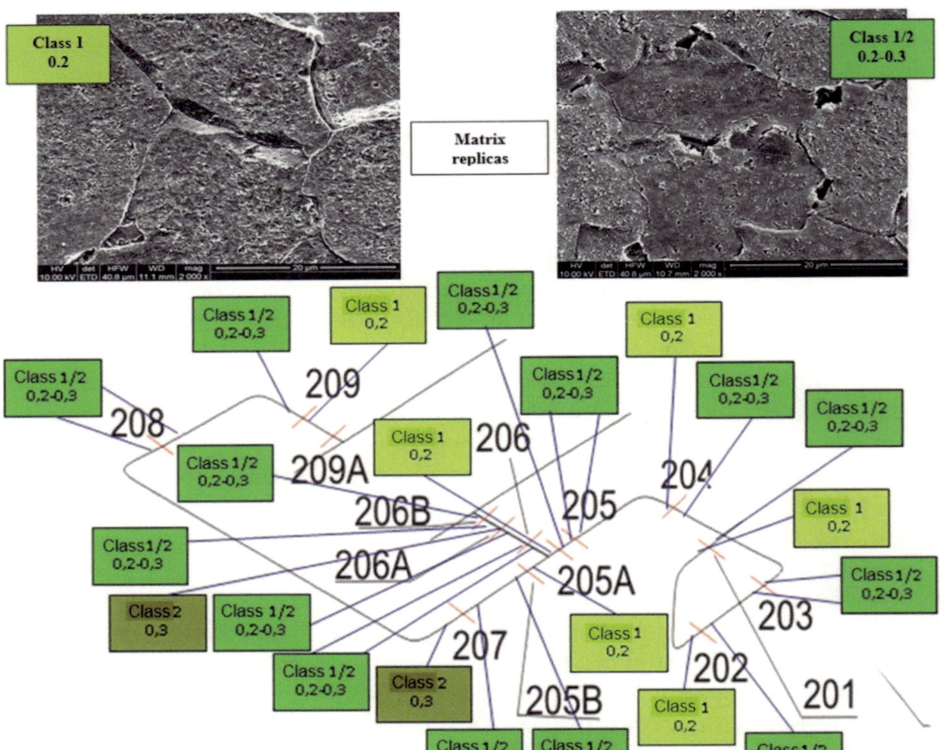

Figure 1.
The class of microstructure and exhaustion degree of the material of straight connection pipes (before the weld at the inlet side and after the weld at the agent outlet side) of the section of 200 MW primary steam pipeline unit boiler after 220 h service under creep conditions.

2. Condition assessment of the material of welded joints in components operated beyond the design time by non-destructive methods

The criterial element deciding about the suitability for further service of butt-welded joints of components operating under creep conditions is the exhaustion degree of basic materials determined based on the revealed state of microstructure and its determined class as well as the degree of internal damages and the class of damages of the heat-affected zone (HAZ) and weld material. The example of the results of non-destructive tests on microstructure of the primary

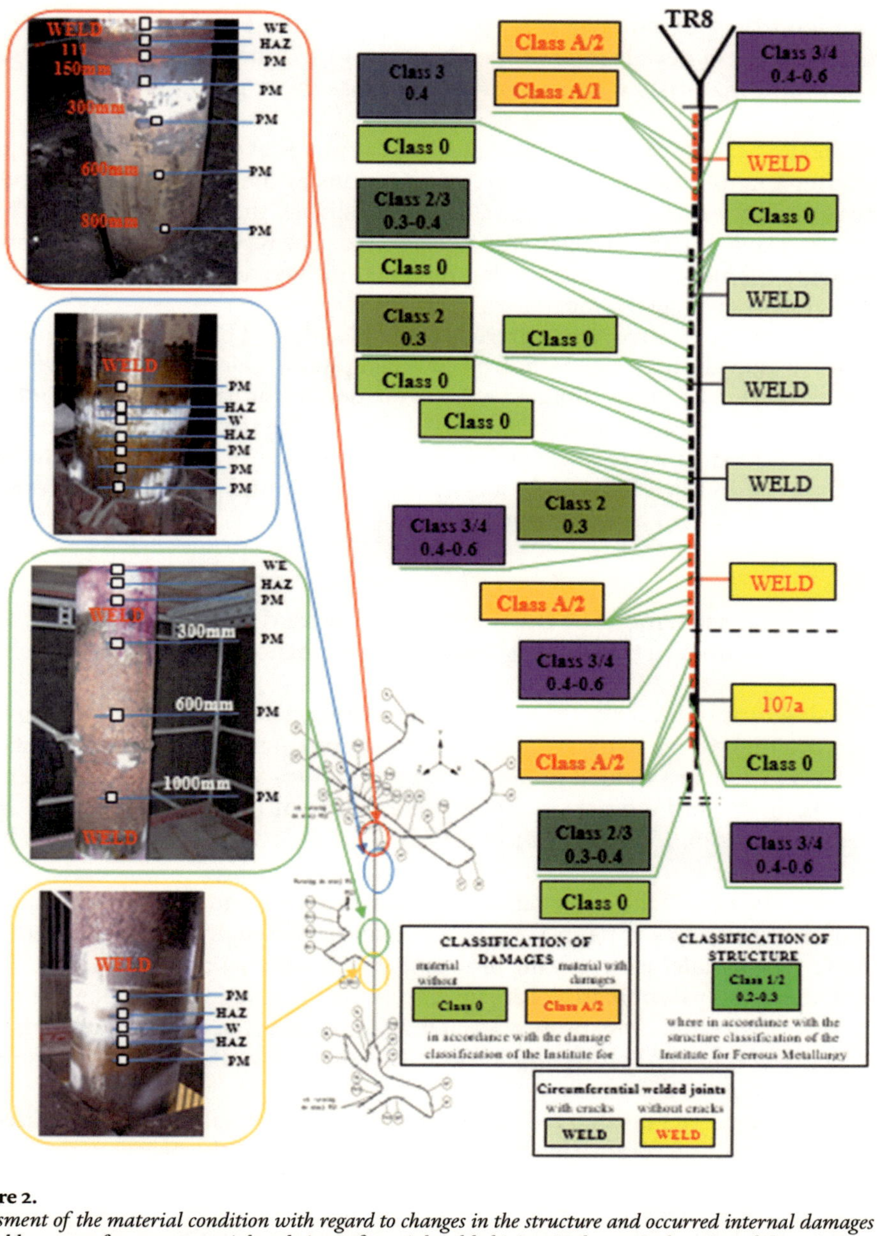

Figure 2.
Assessment of the material condition with regard to changes in the structure and occurred internal damages caused by creep of parent material and circumferential welded joints in the vertical section of the primary steam pipeline made of 14MoV6-3 steel after approximately 220,000 h service (matrix replicas, SEM).

steam pipeline components after long-term service under creep conditions far beyond the design service time without internal damages caused by creep is shown in **Figure 1**, while those with disclosed damages in the form of voids both in the basic material and in the heat-affected zone and weld material are shown in **Figure 2**. The results of these tests allow to take decisions whether the test pipelines can be left and only some of its components need to be replaced using the joining technology by welding materials with significant changes in microstructure caused by creep with new materials.

The assessment of the exhaustion degree and the degree of internal damages in the material of welded joint components allows for determination of the residual and disposable life of the welded joint, capability of the materials of welded joint components to make a possible repair to the test joint, capability of the test basic materials to produce a repair (old/old material) or modernisation (old/new material) welded joint and propose the appropriate welding technology.

To ensure that the condition assessment of the material of welded joint components after long-term service under creep conditions by non-destructive methods is accurate and the estimated exhaustion degree is reliable, the results of destructive materials testing of these components are required. The results of these tests should allow the assignment of exhaustion degree to determined degrees of microstructural degradation of such joints.

The material for destructive materials testing is selected based on the results obtained by non-destructive method of matrix replicas, which determine the class of microstructure and the estimated exhaustion degree for parent material and the class of damage for butt components of welded joints. The material for destructive testing in the form of a test specimen of the component with butt-welded joint is taken from the areas with the highest exhaustion degree and adopted as representative for the component or installation.

3. Condition assessment of the material of welded joints operated beyond the design time using destructive methods

The destructive materials testing on long-time operated welded joints are carried out on homogeneous butt-welded joints of thick-walled components. Most often, these are welded joints of steam pipelines as well as steam superheater chambers and temperature controllers made of the following steel grades: 13CrMo4-5, 10CrMo9-10, 14MoV6-3, X20CrMoV121 and sometimes 15Mo3.

These tests include the observations of microstructure and hardness measurement of different areas of joints from parent material through heat-affected zone and weld. The simultaneous testing of mechanical properties and residual life of the welded joints and parent material are used to search for relations between the state of microstructure and these properties. It is also important to search for correlation between the mechanical properties, especially residual life, of the parent material and welded joint.

The characteristics created based on the obtained test results are used for assessing the condition and forecasting the further safe service of the parent material and welded joints operated beyond the design service time. Moreover, they are decisive in making decisions on the possibility of repairing welded joints using the materials after long-term service under creep conditions and developing the joining technologies.

3.1 Structure assessment of the material of welded joints operated for a long time under creep conditions

The selected example showing the state of structure of homogeneous butt-welded joints after long-term service compared to the structure of the parent material of a component after service far beyond the design service life is shown in **Figure 3** and summarised in **Table 1**.

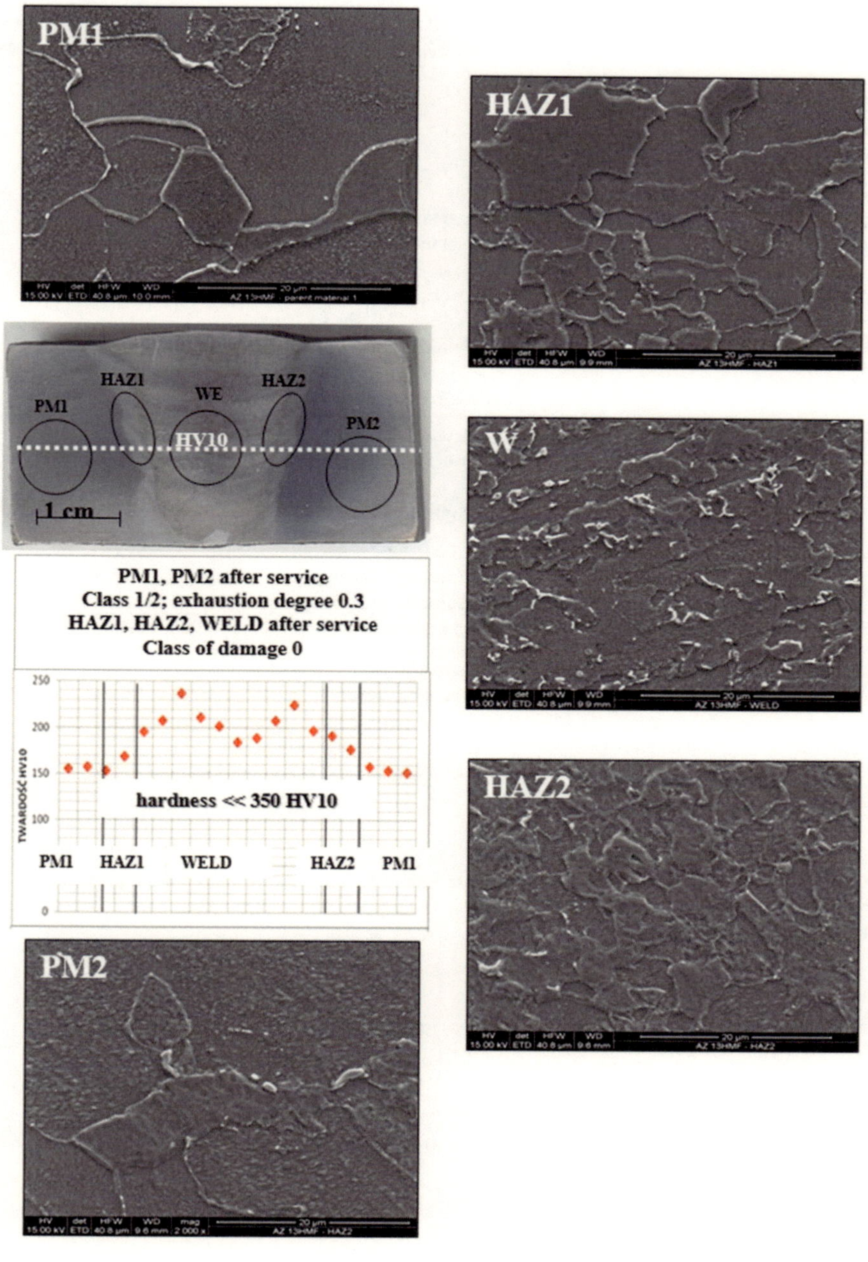

Figure 3.
Structure of the material of homogeneous butt-welded joint components of the main primary steam pipeline made of 14MoV6-3 steel after approximately 200,000 h service under creep conditions parent material marked PM1, PM2; weld material marked WELD heat-affected zone marked HAZ1, HAZ2.

Figure 3 reveals the images of microstructure and the results of hardness measurements of components of the butt-welded joint in the main primary steam pipeline, whereas the description, state of microstructure of the parent material and its exhaustion degree as well as the class of internal damages of the heat-affected zones and weld for the selected example of welded joint are summarised in **Table 1**. The image of microstructure of the welded joint components after service, exceeding

Material used in testing		**Description of microstructure Material condition—exhaustion degree**	**Hardness HV10**
Component Grade of material Service time	**Test area**		
Primary steam pipeline 14MoV6-3 (13HMF) 200,000 h	Parent material **marked PM1**	Ferritic-bainitic structure. Partially coagulated bainitic areas. Few precipitates of various sizes at the ferrite grain boundaries. Precipitates inside ferrite grains, mostly very fine, distributed evenly No discontinuities and micro-cracks are observed in the structure Bainitic areas: class I, precipitates: class a Damaging processes: class O **Class 1/2, exhaustion degree: approximately 0.3**	156
	Parent material **marked PM2**		153
	Heat-affected zone **marked HAZ1**	Bainite with a small amount of ferrite. Significant number of fine precipitates in the bainitic areas and ferrite No discontinuities and micro-cracks are found in the structure	198
	Weld **marked WELD**	Bainitic areas of various sizes with a small number of ferrite. Bainitic areas with precipitates of various sizes No discontinuities and micro-cracks found in the structure	211
	Heat-affected zone **marked HAZ2**	Mixture of bainite and ferrite. Fine precipitates: in the bainite areas—distributed unevenly, in ferrite—distributed evenly. No discontinuities and micro-cracks found in the structure	181

Table 1.
State of microstructure, degree of exhaustion and hardness of the material of homogeneous butt joint components of test critical pressure components of power units after long-term service under creep conditions beyond the design service time, based on examples.

Standard	**Chemical composition, %**								
	C	**Mn**	**Si**	**P**	**S**	**Cr**	**Mo**	**V**	**Ni**
PN-75/H-84024	0.10	0.40	0.15	Max	Max	0.30	0.50	0.22	Max
Cr-Mo-V (13HMF)	0.18	0.70	0.35	0.040	0.040	0.60	0.65	0.35	0.30
Examined element MR1	0.16	0.61	0.33	0.020	0.020	0.48	0.51	0.31	0.08
Examined element MR 2	0.14	0.51	0.29	0.018	0.019	0.52	0.51	0.28	0.09

Table 2.
Analysis of chemical composition.

twice the value of the design service time, most often corresponds to the exhaustion degree which allows the material to be used in further service. It mainly results from significant difference between the design thickness and the actual thickness of the assessed components ($g_o < g_{act}$), which causes that the actual operating stress is lower than the design one ($\sigma_{act} < \sigma$), the as-received condition of the test materials and the correct previous service. Also, the frequent small differences in hardness level in the individual areas of the joints are a positive phenomenon affecting the suitability for further service.

Table 2 presents an analysis of the chemical composition of the tested sections. The mechanical properties of the material obtained from pipeline sections were shown in graphical form in **Figures 4–8**. In order to assess the properties of the tested steels and their welded joints, reference was made to the requirements contained in PN-75/H-84024.

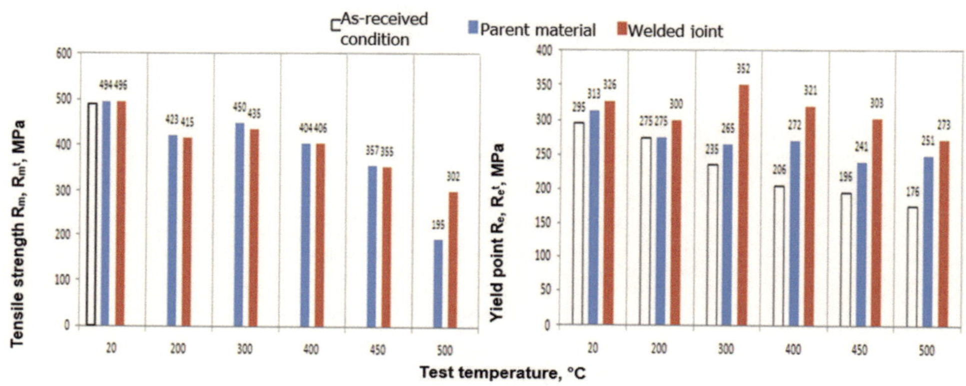

Figure 4.
The comparison of test results for yield point at room (R_e) and elevated (R_e^t) temperature of the parent material and butt-welded joint of the primary steam pipeline made of 14MoV6-3 steel after 240,000 h service under creep conditions in relation to the as-received condition of the material.

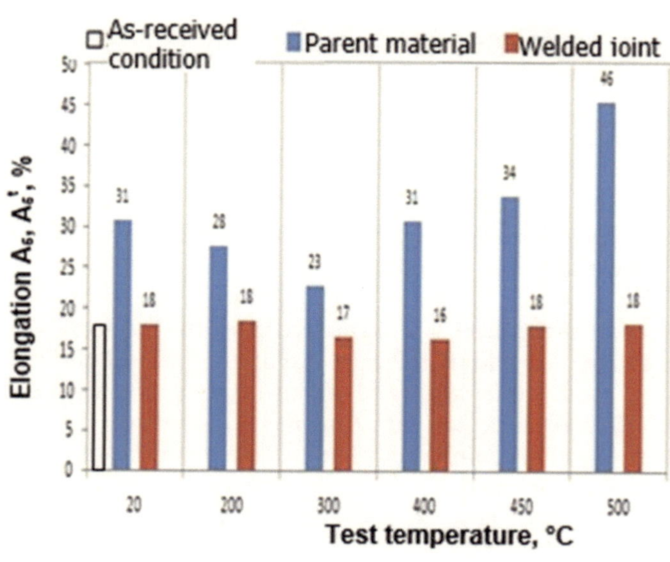

Figure 5.
The comparison of test results for A_{5t} elongation at room and elevated temperature of the shell material and butt-welded joint of the primary steam pipeline made of 14MoV6-3 steel after 240,000 h service.

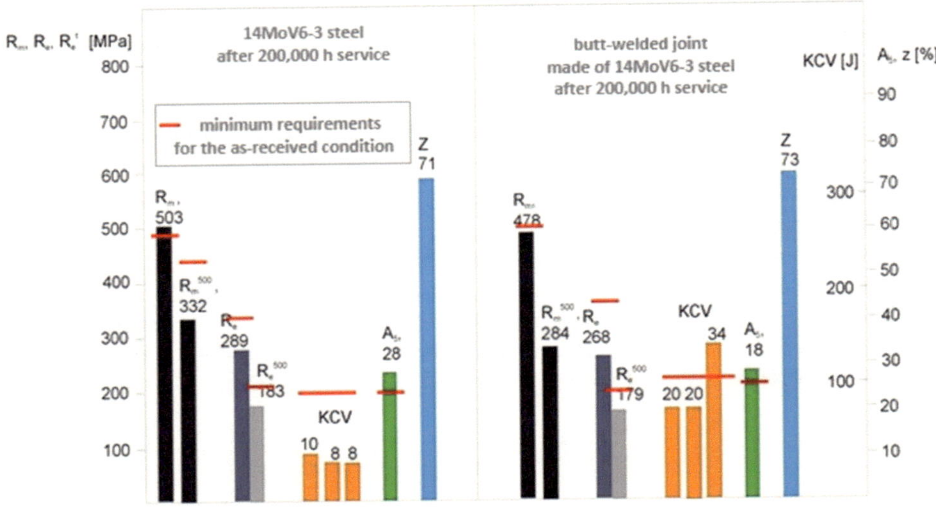

Figure 6.
The comparison of test results for mechanical properties of the parent material and homogeneous butt-welded joint of pipeline components made of 14MoV6-3 after long-term service under creep conditions for 200,000 h.

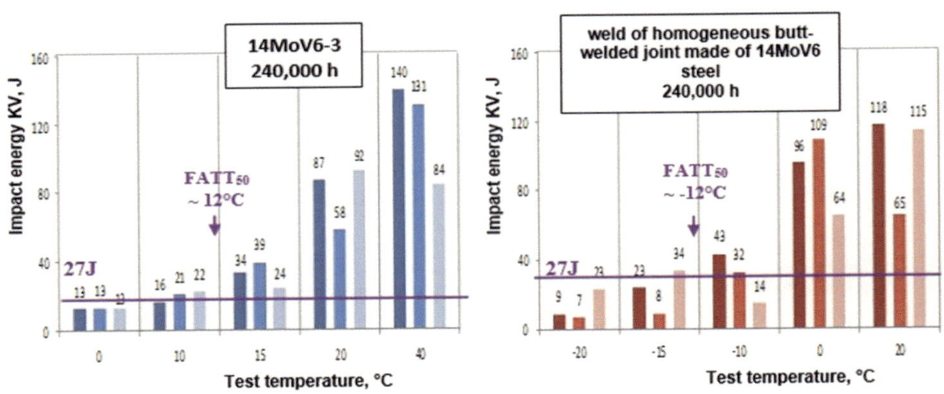

Figure 7.
The comparison of test results for impact energy measured on V-notch test specimens depending on test temperature of parent material and weld of homogeneous butt-welded joint of pipeline components after long-term service under creep conditions: (a) made of 14MoV6-3 steel, (b) made of 10CrMo9-10 steel.

3.2 Assessment of mechanical properties of the material of welded joints operated for a long time under creep conditions

Another element of the materials characteristics is the mechanical properties of welded joints after service and their level. Most often, the level of tensile strength and yield point at room temperature and of tensile strength at elevated temperature of such welded joints after service, for a time much longer than the design time, corresponds to the required minimum values for the parent material in the as-received condition, or their actual values are not much lower. The example of such characteristics is shown in **Figure 4**. The level of these properties is very important for the determination of the rate of start-ups and shutdowns, which is connected with the level of instantaneous internal stresses resulting from the temperature gradient, particularly of thick-walled components. It is also important to know the formability of the material of welded joints, which can be reflected by the

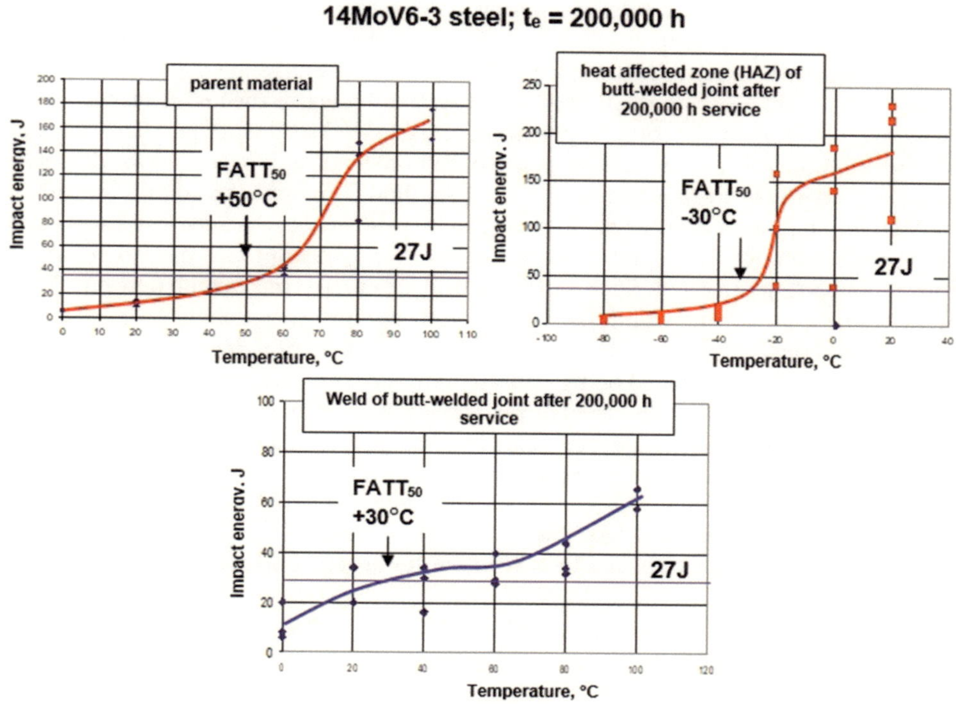

BRITTLE FRACTURE APPEARANCE TRANSITION TEMPERATURE, °C		
+50	-30	+30
parent material after 200,000 h service	Heat-affected zone (HAZ)	weld
	butt-welded joint after 200,000 h service	

Figure 8.
The relationship between the impact energy measured on V-notch test specimens and the test temperature of butt-welded joint components of steam pipeline made of 14MoV6-3 steel after 200,000 h service and the brittle fracture appearance transition temperature determined for (a) parent material, (b) heat-affected zone and (c) weld.

elongation in tensile test, its impact energy and brittle fracture appearance transition temperature both in the weld and the heat-affected zones. The example of elongation test results depending on the test temperature of welded joints compared to the parent material after long-term service far beyond the design service time is shown in **Figure 5**, whereas **Figure 8** presents the comparison of mechanical properties of the parent material and butt-welded joint with regard to the requirements for the parent material in the as-received condition.

The test results for impact energy measured on V-notch test specimens depending on the test temperature of parent material and weld of homogeneous butt-welded joint of the pipeline components after long-term service under creep conditions are compared in **Figure 7**, whereas **Figure 8** shows the relationship between the impact energy measured on V-notch test specimens and the test temperature of the parent material, heat-affected zone and weld of the butt-welded joint of the main steam pipeline made of 14MoV6-3 steel after 20,000 h service as well as brittle fracture appearance transition temperatures determined for the materials of these areas. This example illustrates how diversified the level of brittle fracture appearance transition temperature of the material of individual welded joint components after long-term service under creep conditions can be.

In accordance with the results of long-term own research, for the materials that most often exist in the components operated under creep conditions far beyond the design service time, such as 13CrMo4-5, 14MoV6-3, 10CrMo9-10 low-alloy steels and X20CrMoV121 high-chromium steel, the highest trend towards the loss of plastic properties is shown by the components made of 14MoV6-3 steel and then those made of X20CrMoV121 steel. The knowledge of these material features is necessary for the selection of water-pressure test parameters and the method for overhaul works and repairs.

It should be noted that for components in long-term use, the leak water tests can only be carried out. The use of strength water test may result in destruction of the component being operated. Such a test can only be carried out for the new equipment, and the governing standards apply to new equipment only.

It can be concluded that with an increase in the exhaustion degree, the mechanical properties of the material being operated are reduced. However, no unambiguous relationships between the level of properties, both strength and plastic, and the residual life and exhaustion degree caused by creep were observed so far. According to the present knowledge, such a correlation does not exist.

3.3 Assessment of residual life and disposable residual life of the material of welded joints operated for a long time under creep conditions

The most critical element of the materials characteristics is the creep resistance level of the welded joints being operated. This creep resistance is responsible for

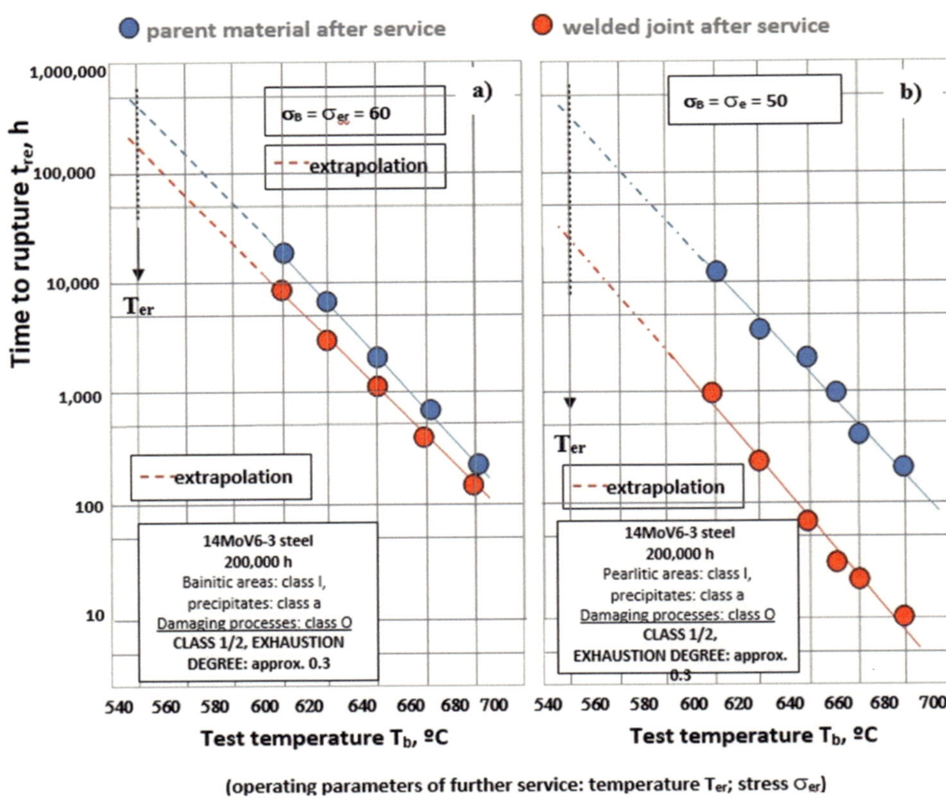

Figure 9.
Residual life of parent material and butt-welded joints of components working far beyond the design service time based on results of short-term creep tests in the form of log $t_{re} = f(t_b)$ *at* σ_b = *const for two different main primary steam pipelines made of 14MoV6-3 steel after 200,000 h service.* σ_B*- test stress.*

Steel grade	Test area	Material used in testing		Adopted operating parameters of further service		Determined residual life, h	Estimated disposable residual life, h
		Type of component Service time	Dimensions of $D_{out} \times g_n$, mm	Stress σ_D MPa	temperature T_D, °C		
14MoV6-3	Parent material	Primary steam pipeline 200,000 h	ϕ327 × 40	60	540	**300,000**	**165,000**
	Welded joint					**200,000**	**110,000**
14MoV6-3	Parent material	Primary steam pipeline 200,000 h	ϕ327 × 40	50	540	**140,000**	**77,000**
	Welded joint					**35,000**	**19,000**

Table 3.
Residual life and disposable residual life based on the results of short-term creep tests on the materials of the main primary steam pipelines made of 14MoV6-3 steel after long-term service far beyond the design service time.

their ability to transfer the actual service loads of the component, which occur under operating conditions. A measure of creep resistance is the level of residual creep resistance, which is able to meet these expectations. However, in the engineering practice, the safe service of the welded joints operated under the preset temperature and pressure conditions is important. The time of safe service is the disposable residual life, which is part of the residual life being determined.

The examples of residual life, which is the time to rupture under the temperature and stress conditions equal to operating ones, determined in short-term creep tests and compared to the residual life of the parent material after service are shown in **Figure 9**.

The determined residual life and disposal residual life for the presented examples are summarised in **Table 3**. The comparison of the residual life and disposable residual life determined for the material of two primary steam pipelines made of 14MoV6-3 steel after 200,000 h service shows significant differences. These differences probably result mainly from differences in the as-received condition of the operated material and welded joint as well as different service history. This confirms the need of an individual assessment, mainly based on the assessment of the actual condition of the component and its material.

4. Summary

Based on many power units in use, the results of creep tests on the butt-welded joints after long-term service beyond the design time carried out on the parent material and butt-welded joints of the components operating under creep conditions invalidate the generally accepted principle that the residual creep strength of the material of welded joint is no less than 80% the residual creep strength of the basic material. The principle of the relationship of residual creep strength between them transferred from that applicable to the basic material and welded joint in the as-received condition is not supported by the investigations of real cases. Also, in many investigated cases, the residual life of the welded joint is significantly lower than that obtained for the parent material. Therefore, for the components and materials operated far beyond the design time, it is necessary to determine the residual life not only for the basic material of the representative test specimen of the component but also for the welded joint (**Table 3**).

Author details

Janusz Dobrzański* and Adam Zieliński
Institute for Ferrous Metallurgy, Gliwice, Poland

*Address all correspondence to: jdobrzanski@imz.pl

References

[1] Zieliński A, Sroka M, Dudziak T. Microstructure and mechanical properties of Inconel 740H after long-term service. Materials. 2018;**11**:2130. DOI: 10.3390/ma11112130

[2] Zieliński A, Dobrzański J, Purzyńska H, Sikora R, Dziuba-Kałuża M, Kania Z. Evaluation of creep strength of heterogeneous welded joint in HR6W alloy and Sanicro 25 steel. Archives of Metallurgy and Materials. 2017;**62**(4):2057-2064

[3] Sroka M, Zielinski A, Hernas A, Kania Z, Rozmus R, Tanski T, et al. The effect of long-term impact of elevated temperature on changes in the microstructure of Inconel 740H alloy. Metalurgija. 2017;**56**(3-4):333-336

[4] Sroka M, Nabialek M, Szota M, Zielinski A. The influence of the temperature and ageing time on the NiCr23Co12Mo alloy microstructure. Revista de Chimie. 2017;**68**(4):737-741

[5] Golanski G, Zielinska-Lipiec A, Zielinski A, Sroka M. Effect of long-term service on microstructure and mechanical properties of martensitic 9% Cr steel. Journal of Materials Engineering and Performance. 2017;**26**(3):1101-1107

[6] Zieliński A, Golański G, Sroka M. Influence of long-term ageing on the microstructure and mechanical properties of T24 steel. Materials Science & Engineering, A: Structural Materials: Properties, Microstructure and Processing. 2017;**682**:664-672

[7] Dudziak T, Deodeshmukh V, Backert L, Sobczak N, Witkowska M, Ratuszek W, et al. Phase investigations under steam oxidation process at 800°C for 1000 h of advanced steels and Ni-based alloys. Oxidation of Metals. 2017;**87**(1-2):139-158

[8] Zieliński A, Golański G, Sroka M, Skupień P. Microstructure and mechanical properties of the T23 steel after long-term ageing at elevated temperature. Materials at High Temperatures. 2016;**33**:154-163

[9] Zieliński A, Golański G, Dobrzański J, Sroka M. Creep resistance of VM12 steel. Archives of Metallurgy and Materials. 2016;**61**:1289-1294

[10] Wojsyk K, Golański G, Jasek J, Słania J, Zieliński A, Urbańczyk P. Influence of the annealing time after welding on the mechanical properties of welded joint of T91 steel. Archives of Metallurgy and Materials. 2016;**61**:1079-1084

[11] Zieliński A, Boryczko B, Miczka M, Sroka M. Forecasting in the presence of microstructural changes for the case of P91 steel after long-term ageing. Archives of Civil and Mechanical Engineering. 2016;**16**:813-824

[12] Zieliński A, Sroka M, Miczka M, Śliwa A. Forecasting the particle diameter size distribution in P92 (X10CrWMoVNb9-2) steel after long-term ageing at 600 and 650°C. Archives of Metallurgy and Materials. 2016;**61**:753-760

[13] Zieliński A, Sroka M, Hernas A, Kremzer M. The effect of long-term impact of elevated temperature on changes in microstructure and mechanical properties of HR3C steel. Archives of Metallurgy and Materials. 2016;**61**:761-765

[14] Zieliński A, Purzyńska H, Dobrzański J, Golański G. Changes in properties and structure of high-chromium 9-12% Cr steels due to long-term exposure at elevated temperature. Archives of Metallurgy and Materials. 2016;**61**:957-964

[15] Zieliński A, Paszkowska H, Skupień P, Golański G. Assessment of service

life of HCM12 steel after 100,000 hrs of service. Archives of Metallurgy and Materials. 2016;**61**:1021-1029

[16] Zieliński A, Dobrzański J, Purzyńska H, Golański G. Properties, structure and creep resistance of austenitic steel super 304H. Materials Testing. 2015;**57**:859-865

[17] Golański G, Lis A, Słania J, Zieliński A. Microstructural aspect of long term service of the austenitic TP347HFG steel. Archives of Metallurgy and Materials. 2015;**60**:2901-2904

[18] Golański G, Zieliński A, Zielińska-Lipiec A. Degradation of microstructure and mechanical properties in martensitic cast steel after ageing. Materialwissenschaft und Werkstofftechnik. 2015;**46**(3):248-255

[19] Golański G, Zieliński A, Słania J, Jasak J. Mechanical properties of VM12 steel after 30000 hrs of ageing at 600°C temperature. Archives of Metallurgy and Materials. 2014;**59**(4):1351-1354

[20] Kępa J, Golański G, Zieliński A, Brodziak-Hyska A. Precipitation process in VM12 steel after ageing at 650°C temperature. JVE - Journal of Vibroengineering. 2012;**14**(1):143-150

[21] Dobrzański J, Hernas A. Correlation between phase composition and life-time of 1Cr-0.5Mo steels during long term service at elevated temperature. Journal of Materials Processing Technology. 1995;**53**:101-108

[22] Dobrzański J, Hernas A. Relationship between microstructure and remanent life-time of low alloy Cr-Mo steels. In: 6th International Conference on Creep and Fatigue. London; 1996. p. 451

[23] Dobrzański J. The classification method and the technical condition evaluation of the critical elements' material of power boilers in creep service made from the 12Cr-1Mo-V. Journal of Materials Processing Technology. 2005;**164-165**:785-794

[24] Neubauer B. Creep damage evolution in power plants. In: Proceeding of the Second International Conference on Creep and Fracture of Engineering Materials and Structures. Swansea, UK; 1984. pp. 1226-1271

[25] Dobrzański J. Internal damage processes in low alloy chromium-molibdenum steels during high-temperature creep service. Journal of Materials Processing Technology. 2004;**157-158**:297-303

[26] Dobrzański J, Sroka M, Zieliński A. Methodology of classification of internal damage the steels during creep service. Journal of Achievements in Materials and Manufacturing Engineering. 2006;**18**:263-266

[27] Dobrzański J, Sroka M. Computer aided classification of internal damages the chromium-molybdenum steels during creep service. Journal of Achievements in Materials and Manufacturing Engineering. 2007;**24**(3):143-146

[28] Zieliński A, Dobrzański J, Krztoń H. Structural changes in low alloy cast steel Cr-Mo-V after long time creep service. Journal of Achievements in Materials and Manufacturing Engineering. 2007;**25**(1): 33-36

[29] Dobrzański J, Krztoń H, Zieliński A. Development of the precipitation processes in low-alloy Cr-Mo type steel for evolution of the material state after exceeding the assessed lifetime. Journal of Achievements in Materials and Manufacturing Engineering. 2007;**23**(2):19-22

[30] Dobrzański J, Zieliński A, Krztoń H. Mechanical properties and structure of the Cr-Mo-V low-alloyed steel after

long-term service in creep condition. Journal of Achievements in Materials and Manufacturing Engineering. 2007;**23**(1):39-42

[31] Zieliński A. The effect of long-term service under creep conditions on mechanical properties of Cr-Mo-V cast steel. In: 8th International Scientific Conference—New Technologies and Achievements in Metallurgy and Materials Engineering. Vol. II. Czestochowa: Publishing House of the Czestochowa University of Technology; 2007. pp. 717-721

[32] Hernas A, Dobrzański J. Life Time and Destruction of Steam Boilers and Turbines. Gliwice: The Silesian University of Technology Publishing House; 2003

[33] Dobrzański J, Hernas A, Pasternak J, Zieliński A. Microstructure and mechanical properties characteristics of welded joints made of creep-resistant steel with 12% Cr, V, W and Co additions. In: Advances in Materials Technology for Fossil Power Plants. ASM International; 2008. pp. 303-319. DOI: 10.1361/cp2007epri0303

[34] Zieliński A, Dobrzański J, Renowicz D, Hernas A. The estimation of residual life of low-alloy cast steel Cr-Mo-V type after long-term creep service. In: Advances in Materials Technology for Fossil Power Plants. ASM International; 2008. pp. 616-626. DOI: 10.1361/cp2007epri0616

Chapter 5

The Procedure for Determining the Time of Safe Service beyond the Design Service Time Based on Creep Testing

Janusz Dobrzański

Abstract

This article describes the method for determination of residual life and disposable residual life based on long- and short-term creep tests for different temperature levels above the operating one and stress level equal to the operating one. The method for determination of the share of disposable life in total life and the time of safe service beyond the design service time was proposed. These characteristics of the steel are used for the evaluation of the structural changes and mechanical properties of the material after long-term operation. The result of this study is the database of material characteristics representing the mechanical properties related to the structure analysis, and it can be used for diagnosis of the components of pressure part of power boilers.

Keywords: microstructure, mechanical properties, short creep test, residual life, welded joint

1. Introduction

The extension in the service time of power units which have exceeded the design service life of 100,000 h is made using the calculation method based on the creep strength data for 200,000 h service and the positive results of comprehensive diagnostic tests, with particular consideration given to those of the critical pressure components of boilers and turbines. Out of them, the components working at above the limit temperature, that is, under creep conditions, are of particular importance. In the assessment of these components, it is important and necessary to assess the condition of their materials [1–7]. Such an assessment is carried out based on non-destructive materials testing, and its result is referred to the available characteristics of materials after service [8–15]. The results of these tests provide good estimation of the condition of material and its exhaustion degree and allow for determination of the time of further safe service until the next inspection [16–28].

However, to reach the level of 200,000 h service for the material of components, not only good estimation of residual life but also its determination based on the destructive testing on a representative test specimen is required in a number of cases.

Yet, it is not always possible in practice. It can be made to assess the condition of the material of the main steam pipeline or transfer pipeline as well as for certain chambers and desuperheaters. However, it must be preceded by a cost-effectiveness calculation for such a procedure.

The problem related to the assessment of creep resistance in case of materials operated for a long time under creep conditions is the time required to carry out the creep tests to assess the residual life and the remaining safe time of service. This paper shows how to use long-term creep tests at the temperature and stress parameters corresponding to the actual operating ones, which allow for determination of the actual residual life, to make such an assessment. The method for using short-term creep tests for such an assessment where test duration is reduced by increasing one of the parameters in relation to the actual ones is also presented.

2. Creep tests

2.1 Long-term creep tests

In the assessment of residual life and disposable residual life, the main source of information is still creep tests in spite of the fact that their disadvantage is a long time until the test results are obtained. Creep tests are usually carried out under uniaxial tension and at strain rates from 10 to 9 to 10–12 s^{-1} (corresponding to the actual service conditions) on test specimens taken directly from the material of component after long-term service at a temperature exceeding the limit one tg (te > tg).

Long-term creep tests can be divided into:

- Creep rupture tests without measurement of elongation during test
- Creep test with measurement of elongation during test

Based on the results of creep rupture tests without measurement of elongation during test, the residual life can be determined using one of the two procedures:

- Lifetime fracture rule
- Extrapolation of creep rupture test results

The lifetime fracture rule relies on the assumption that changes in temperature T or stress σ during service can be divided into degrees characterised by specific temperature Ti and stress σ_i.

Thus, the lifetime fracture rule concept by Robinson assumes that

$$\sum_i \left[\frac{t(\sigma, T)}{t_r(\sigma, T)}\right] = 1$$

where T is the time of stress σ at a temperature T, tr is the time to rupture under stress σ and at temperature T, and i is the number of degrees of stress and temperature T from the beginning of service until rupture of the material.

For each of these degrees, the creep rate in steady state (secondary creep) or the time to rupture tri based on the experimental data measured under preset stress and temperature is determined. The total creep strain is assumed to be the sum of strains corresponding to the consecutive degrees, that is, $\varepsilon_c = \sum_i \varepsilon_i$.

For the materials after service, the lifetime fracture rule can be presented as

$$\frac{t_s}{t_{r_s}} + \frac{t_t}{t_{r_t}} = 1$$

where t_s is the time of previous service of the material, t_{rs} is the time to rupture under service conditions (σ_e, Te), t_t is the time to rupture of the test specimen determined under the accelerated creep test conditions for the material after service ($\sigma_t > \sigma_e$ and/or Tt > Te), and trt is the time to rupture of the material in the as-received condition under the accelerated test conditions ($\sigma_e = \sigma_t$, T = Tt).

Assuming that the further operating conditions will not differ from the previous ones, the time equal to the residual life can be determined from

$$t_{r_s} - t_s = \frac{t_t t_s}{t_{r_t} - t_t}$$

The need to know the time to rupture of the material in the as-received condition t_{rt} restricts the use of this method.

In the event when the test specimens for destructive testing are taken from the operated material after two different times of service ts1 and ts2 and the creep tests are carried out for both conditions of one material, the time to rupture of the material in the as-received condition trt can be eliminated. Then, the knowledge of the properties of material in the as-received condition is not required.

The use of this method is particularly useful in the modernisation of boilers when some critical boiler components operating under creep conditions are replaced and the basic operating parameters are recorded and controlled on-line.

The residual life can also be determined by extrapolating the results of creep rupture tests carried out on material after service at a temperature equal to or different from the design service temperature (Tr) and stress values equal to, but usually higher than, those characteristic of the service conditions ($\boldsymbol{\sigma_r}$). Obtained in this way, the results of creep rupture tests with a time of usually approx. 10,000 h are presented in the parametric way by expressing one variable as a product of the function of two other variables, that is, stress and temperature and time, respectively, using the Larson-Miller parameter in the following form:

$$H\ (\boldsymbol{\sigma})LM = T\ (C + \log\ t_{re})$$

where H ($\boldsymbol{\sigma}$) is the stress function, $\boldsymbol{\sigma}$ is the stress, T is the temperature in K, C is the material constant at 1/T = 0, and t_{re} is the time to rupture of material after service.

This method for determination of residual life is shown graphically in **Figure 1**. Knowing the operating temperature T_r and the operating stress $\boldsymbol{\sigma_r}$ (resulting from the component's geometry and operating pressure p_r) of further service, the time to rupture t_{re} can be determined following which the component destruction should be expected. However, the determined time t_{re} is of no practical value. The practical value is the time to end of secondary creep, which is called disposable residual life t_{re} 0.6 and which is part of the residual life. This value is specific to each of the test grades of material.

The duration of the tests allowing for determination of the residual life t_{re} can be reduced by carrying out the creep tests with measurement of elongation during test in order to determine the creep rate in steady state for the properly selected creep parameters ($\boldsymbol{\sigma}$b, Tb). Using the obtained curve of maximum creep rate vs. stress = f($\boldsymbol{\sigma}$) at the constant temperature Tep of further service, the time of further safe service can be determined, expressed by the following formula:

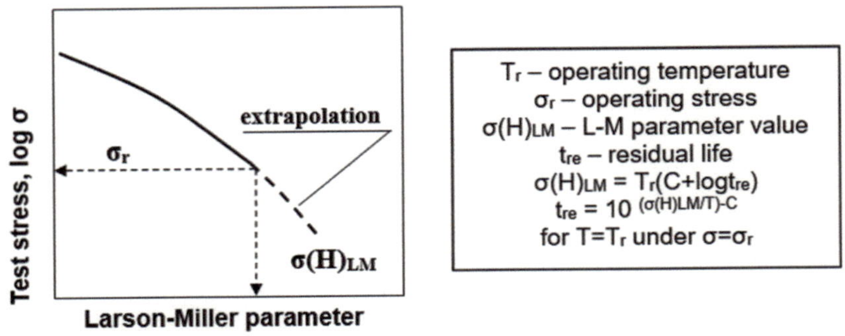

Figure 1.
Determination of residual life using extrapolated Larson-Miller parameter curve for creep strength of the material after service at above limit temperature.

$$t_{e_p} = \frac{\varepsilon_{c_{dop}}}{\varepsilon^{\cdot sep}}$$

where $\varepsilon_{c\ acpt}$ is the acceptable total strain (adopted based on the operating experience and creep test results)—creep rate for parameters of further service (ε_e, Te) provided that the relation $\varepsilon_e \le \varepsilon_{c\ acpt}$ is met.

The method for determination of creep rate for the parameters of further service is shown graphically in **Figure 2**.

The creep rupture tests with measurement of elongation during test are also used to determine:

- The time to end of secondary creep t_{II} and the corresponding elongation ε_{II}
- The time to rupture t_r and the corresponding elongation ε_c
- The share υ of the time of secondary creep t_{II} in the overall time to rupture t_r $\left(\upsilon = \frac{t_{II}}{t_r}\right)$

The tests are carried out at a constant test temperature T_b, which is similar to the operating one, and at stress σ_b values higher than those corresponding to the operating one. As a result of each of the tests, the creep curves $\varepsilon = f(t)$ at σ_b = const are obtained (**Figure 3a**). The test results are plotted on the diagram in the coordinate system $\upsilon = f(\sigma)\varepsilon_c = f(\sigma)\varepsilon_{II} = f(\sigma)$ at $T_b \approx T_e$ = const (**Figure 3b**). The characteristics obtained in this way allow for determination of the disposable residual life, which is the time of further safe service for the adopted operating parameters.

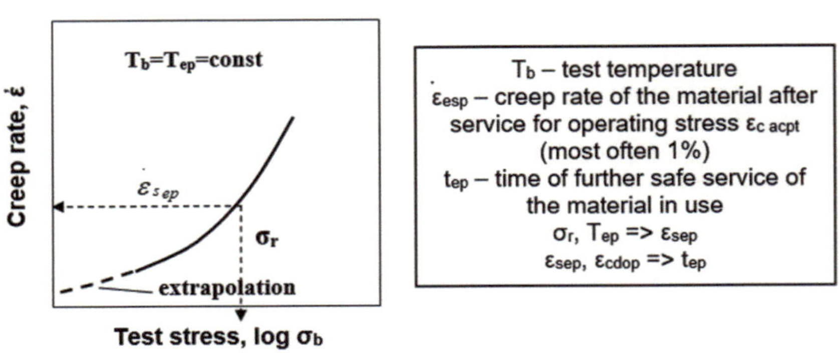

Figure 2.
Relationship between creep rate $\varepsilon^{\cdot}_s$ in steady state of the material after service under creep conditions and test stress for determination of the time of further safe service t_{ep}.

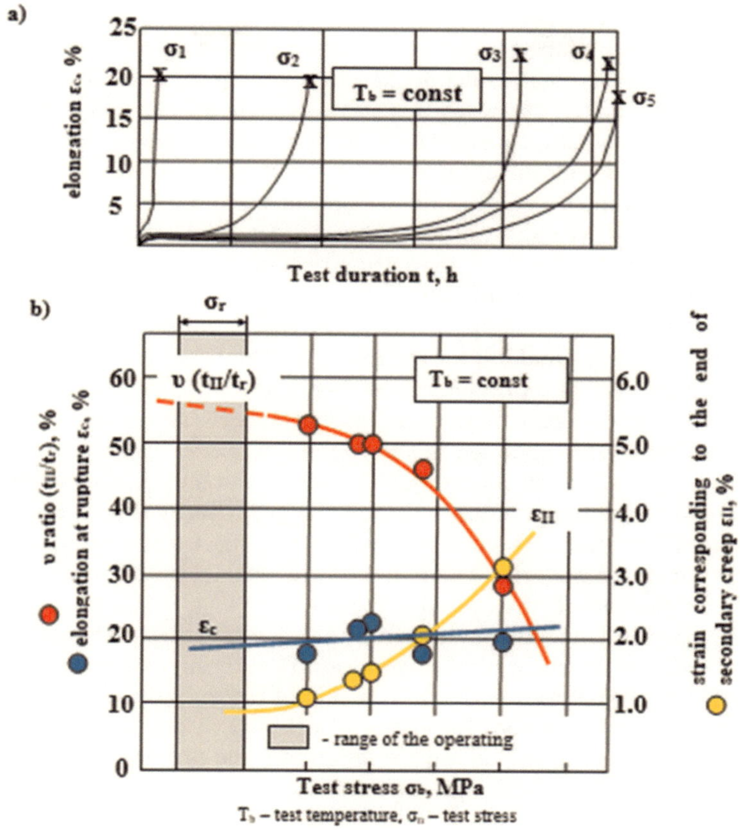

Figure 3.
The results of creep rupture tests with recording of elongation. (a) Creep curves at constant temperature T_b = const. (b) Strain corresponding to the end of secondary creep, elongation at rupture, and υ coefficient depending on test stress.

2.2 Short-term creep tests

The disadvantage of the method for determination of residual life using long-term creep tests is the time of waiting for test results, which takes at least 2–3 years. To reduce the duration of these tests and the assessment of residual life, the so-called short-term creep tests with a duration from a few dozen to max 3000–10,000 h are used in the engineering practice. This opens up the possibility for obtaining test results within a maximum of several months and providing good estimation of residual life.

The acceleration of the creep process and the reduction in the duration of tests is obtained in creep tests carried out under uniaxial tension on test specimens taken from the material of a power system component and:

- Under constant test stress equal to the operating stress and at different test temperature levels much higher than the operating temperature
- At constant test temperature equal to the operating temperature and under different test stress levels much higher than the operating stress

$\mathbf{T_b}$, test temperature; $\mathbf{T_r}$, operating temperature; $\boldsymbol{\sigma_b}$, test stress; $\boldsymbol{\sigma_r}$, operating stress; $\mathbf{t_r}$, time to rupture; $\mathbf{R_{z\ av}}$, average temporary creep strength (result of long-term creep tests); $\mathbf{t_{r(r)s}}$, time to rupture for the operating parameters obtained by short-term creep tests as a result of extrapolation; $\mathbf{t_{r(r)}}$, time to rupture for the operating parameters obtained by long-term creep tests of up to 100,000 h.

The manner of presenting results of short-term creep tests conducted under constant stress equal to the operating one (**σ**b = **σ**r = const) for different levels of test temperature Tb and at a constant temperature equal to the operating one (Tb = tr = const) and under different levels of test stress **σ**b as well as the reliability of their estimation is shown graphically in **Figures 4** and 5.

Based on at least 40 years' experience of the Creep Test Laboratory of the Institute for Ferrous Metallurgy in comparing results of short-term creep tests and extrapolating them with results of long-term tests of even up to 100,000 h, they were found to be useful for and applicable to the estimation of life and residual life. The determination of residual life for the material after service does not require knowledge of its history and previous operating conditions, and the only requirement is to define the operating parameters of its further service (σ_{rp}, T_{rp}). Strong convergence of the results obtained in short- and long-term creep tests allows the method of short-term creep tests conducted under constant stress equal to the operating one to be used in practice. A disadvantage of this method is the restriction with regard to the applied test temperature levels arising from precipitation processes that occur in the material. The temperature levels must therefore be selected individually for each grade of material being tested.

Therefore, in the diagnostic practice, only short-term creep tests under constant stress and at different levels of temperature significantly higher than the operating one should be used to determine the residual life of the materials of components operating under creep conditions (pipelines, steam superheater coils). Such tests are recommended for use after exceeding the design service time, which is most often equal to 100,000 h. When the design time is exceeded by 50% (1.5 of t_o), these tests are obligatory.

The reliability of the obtained results of short-term creep tests depends on the requirements for compliance with constant level of preset test temperature T_b over the measurement length of the test specimen throughout the duration of the test. The determination of residual life with an error of no more than 20% in relation to the time determined based on long-term creep tests ensures that short-term creep tests are carried out at the test temperature T_b (much higher than the operating

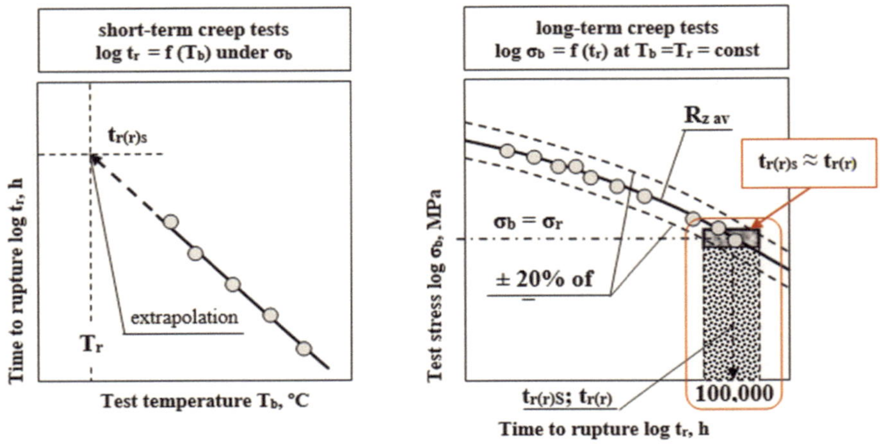

Figure 4.
Reliability of the estimation of time to rupture t_r for the operating parameters based on the results of short-term creep tests under constant test stress equal to the operating one $\sigma_b = \sigma_r$ = const and at different test temperatures T_b higher than the operating one.

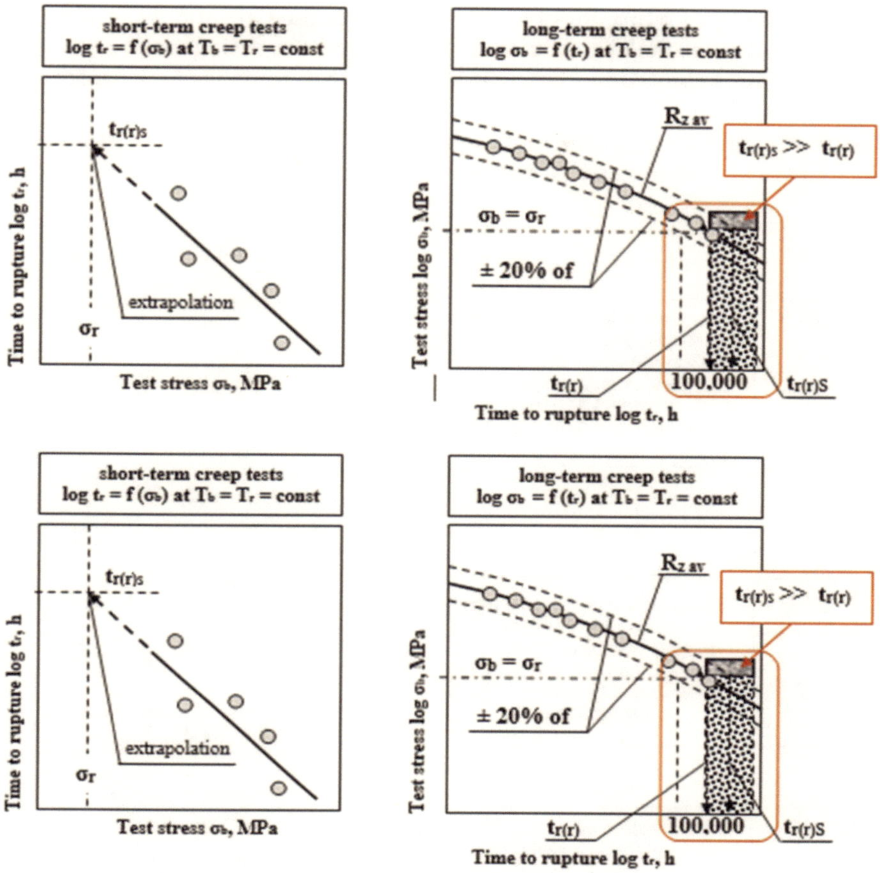

Figure 5.
The reliability of the estimation of time to rupture tr for the operating parameters based on the results of short-term creep tests at a constant temperature equal to the operating one Tb = Tr = const and under different levels of stress σb higher than the operating one. T_b, test temperature; T_r, operating temperature; σ_b, test stress; σ_r, operating stress; t_r, time to rupture; $R_{z\ av}$, average temporary creep strength (result of long-term creep tests); $t_{z(r)s}$, time to rupture for the operating parameters obtained in short-term creep tests as a result of extrapolation; $t_{r(r)}$, time to rupture for the operating parameters obtained in long-term creep tests of up to 100,000 h.

one) with an accuracy equal to or higher than ±0.5°C over the measurement length of the test specimen, regardless of what the test temperature level is.

3. Method for determination of the time of safe service for material after long-term service under creep conditions beyond the design service time

The main step in the proposed method for determination of the time of further safe service is to determine the residual life. Its determination is based on creep test results. These tests have been so far the only known method for determination of service life for actual operating parameters of the materials in use. In creep tests, the decisive factor for their duration is the time to rupture. It cannot be reduced in the development of material characteristics. However, such a possibility exists for the assessment of specific material both in the as-received condition and after service. Nevertheless, the research methods used for this purpose have to be verified with results of long-term creep tests [16].

The proposed procedure for determination of residual life and disposable residual life, which is the time of safe service beyond the design time, is presented as the algorithm presented in **Figure 6**. This algorithm consists of six consecutive steps.

The **first step** is to select the representative areas for destructive testing. At the designated area in the material of test specimen, the short-term creep tests should

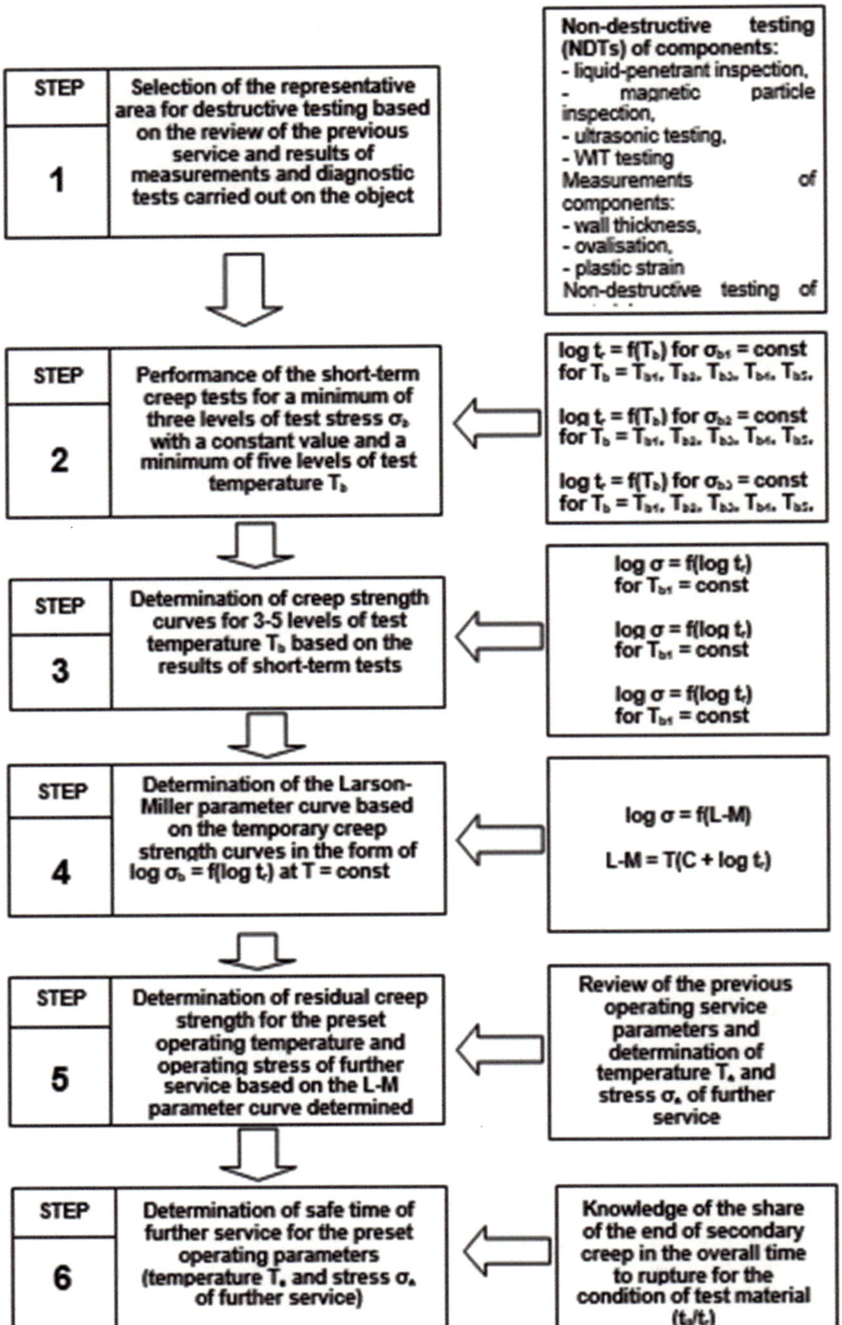

Figure 6.
Procedure for determination of the time of further safe service beyond the design time according to creep strength characteristics based on short-term creep tests under constant stress level.

be carried out under constant stress on at least three different levels (**step 2**). Based on the relationships obtained from the results of short-term creep tests, the curves of temporary creep strength at a constant temperature are determined for a few temperature levels (**step 3**). These creep strength curves allow for determination of the parameter creep strength curve where the parameter is a function of temperature and time to rupture (**step 4**). The obtained parameter creep resistance curve allows the residual creep strength to be determined for the operating parameters (temperature and stress) of further service (**step 5**). When the residual creep strength for the parameters of further service is known, the disposable residual life can be determined. It is the safe time of further service for the adopted operating parameters (**step 6**). The proposed procedure consisting of consecutive steps is presented below on the example of the material after long-term service under creep conditions far beyond the design time.

3.1 Selection of the representative areas for destructive testing and of the test material (step 1)

The selection of the representative areas for destructive testing is made after review of the previous service, results of measurements and diagnostic tests carried out on the object, verification calculations and determination of the temperature, and strain and stress distributions by the finite element method for the most strenuous areas [17–19].

The test materials selected in this way, with provision of the appropriate access to components for cutting them out and the possibility for repair/replacement, were an elbow and a straight section of the main primary steam pipeline made of 13 HMF steel after approx. 200,000 h service.

3.2 Short-term creep tests (step 2)

The short-term creep tests were carried out for six levels of test temperature T_b = 600, 620, 640, 650, 660, and 680°C under the constant level of test stress σ_b with three different values of σ_b = 50, 80, and 100 MPa.

The obtained results of tests on the primary steam pipeline elbow made of 14MoV6–3 steel in the form of $\log t_r = f(T_b)$ under σ_b = const, where t_r is the time to rupture in creep test, for three levels of stress σ_{b1} = 50 MPa, σ_{b2} = 80 MPa, and σ_{b3} = 100 MPa are shown in **Figure 7**.

3.3 Temporary creep strength curves at a constant temperature (step 3)

The extrapolation of the obtained relationships of $\log t_r = f(t_b)$ under σ_b = const towards a lower temperature allowed for determination of residual life t_{re} for temperature equal to the extrapolation one.

The results of this extrapolation for one level of temperature T_b = const and three levels of stress σ_{b1} = 50, σ_{b2} = 80, and σ_{b3} = 100 MPa allowed the creep strength curves to be plotted in the form of $\log \sigma_b = f(\log t_{re})$ for T_{b1} = 530, T_{b2} = 540, T_{b3} = 560, and T_{b4} = 580°C, which is presented graphically in **Figure 8**.

3.4 Parameter residual creep strength curve (step 4)

Based on the constant-temperature residual creep strength curves in the form of $\log \sigma_b = f(\log t_{re})$, the residual creep strength t_{re} was determined, and on its basis, the Larson-Miller parameter curve was determined in the form of $\sigma_b = f(L\text{-}M)$. Where: $L\text{-}M = T_b (C + \log t_{re})$; T_b, test temperature; C, material constant, t_{re}, residual creep strength (**Figure 9**).

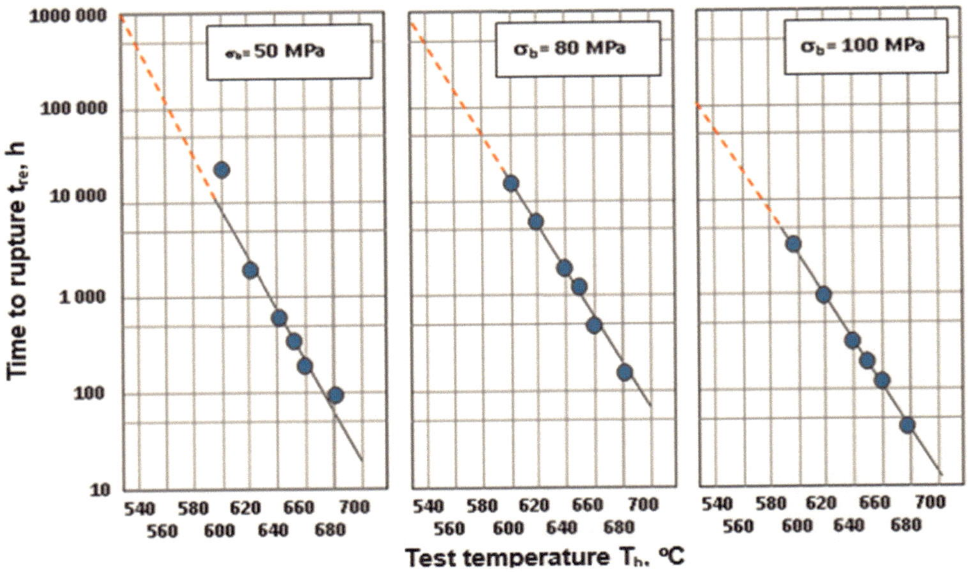

Figure 7.
The results of short-term creep tests on the material of primary steam pipeline elbow made of 13 HMF steel after approx. 200,000 h service under creep conditions in the form of log $t_{re} = f(T_b)$ under σ_b = const.

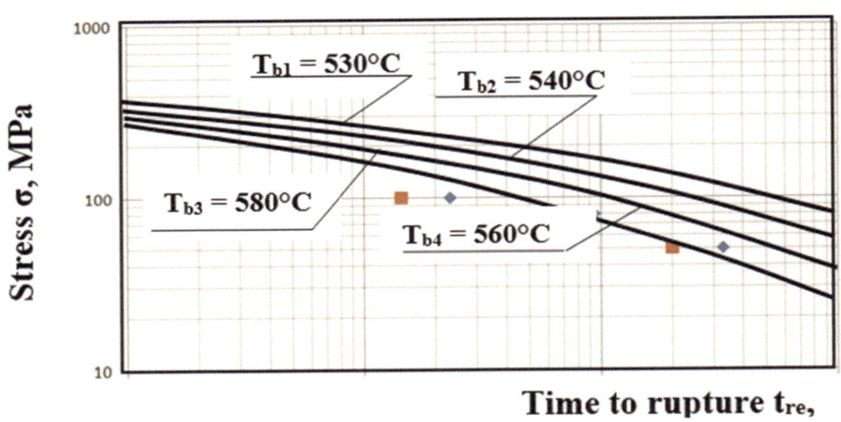

Figure 8.
Residual creep strength curves R_{Ze} determined based on the results obtained from extrapolation of creep tests carried out on test pipeline elbow material made of 13 HMF steel after 200,000 h service under creep conditions.

3.5 Safe time of service beyond the design time (step 5)

In contrast to the actual lifetime equal to the time to rupture of the material subject to creep, which is the residual life t_{re}, the disposable residual life t_{be} is of practical importance. The disposable life is the time until which a structural component can be safely operated under the assumed temperature and stress conditions.

The disposable residual life is part of the residual life. To determine its share in the residual life, defined as the time to the end of the secondary creep t_{IIe} in relation to the time to rupture t_{re}, the rupture creep tests with measurement of elongation during test at a constant temperature T_b = 500°C similar to the operating one were carried out.

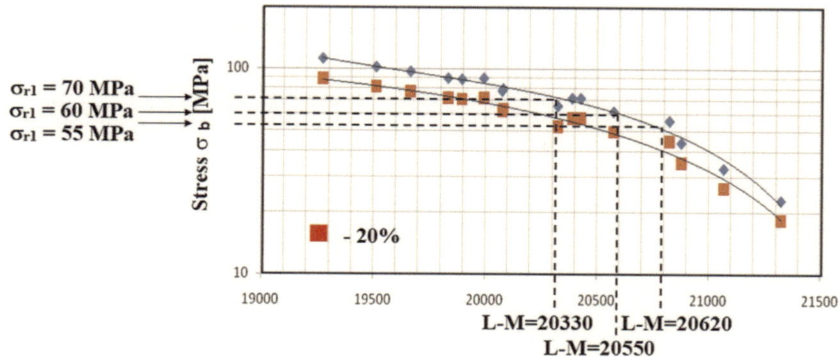

Figure 9.
Residual life t_{re} of the material of the test primary steam pipeline elbow made of 13 HMF steel after approx. 200,000 h service under creep conditions in the form of Larson-Miller parameter curve.

The results of the creep tests shown as creep curves $\varepsilon = f(t)$ at T_b = const and the ratio of the disposable residual life t_{be} to the residual life t_{re} depending on the test stress σ_b allowed to determine the t_{be}/t_{re} ratio for the stress level equal to the operating stress $\sigma_b = \sigma_{ep}$ = 50–60 MPa, which is written as $t_{be} = 0.55t_{re}$ (**Figure 3**).

By knowing the residual life, the disposable residual life can be determined as the time of safe service for the adopted operating parameters. For the material of elbow, the relationships between the disposable residual life t_{be} and the operating temperature of further service T_{ep} for the adopted constant level of operating stress σ_{ep} at the selected levels of operating temperature T_{ep} = 540 and 550°C and under stress $\sigma_b = \sigma_{ep}$ = 60 and 70 MPa are shown in **Figure 10a** and for the circumferential welded joint at T_{ep} = 530 and 540°C and operating stress σ_{ep} = 50 MPa in **Figure 6**, whereas the safe service time of the material of test primary steam pipeline elbow and circumferential welded joint made of 13 HMF steel after approx. 200,000 h service under creep conditions for the operating parameters of further service, which was determined based on the above-mentioned relationships, is summarised in **Table 1**.

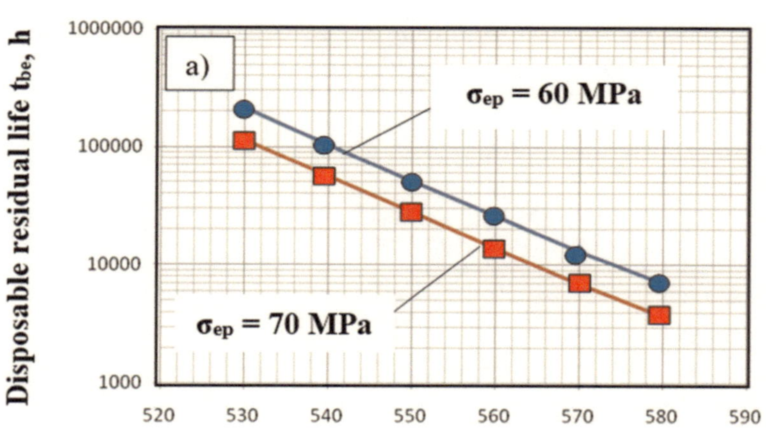

Figure 10.
Disposable residual life t_{be} of the material of primary steam pipeline elbow made of 13 HMF steel after 200,000 h service under creep conditions and stress σ_{ep} = 60 and 70 MPa equal to the operating one depending on operating temperature T_{ep}.

Component	Operating temperature T_{ep}, °C	Operating stress σ_{ep}, MPa		
		55	60	70
		Safe time of further service t_{be}, h		
Elbow	540	127,000	104,000	56,000
	550	62,000	51,000	28,000
Circumferential welded joint	**Operating temperature T_{ep}, °C**	**Operating stress σ_{ep}, MPa**		
		50		80
		Safe time of further service t_{be}, h		
	530	143,000		48,400
	540	77,000		26,000

Table 1.
Safe service time of the material of the test primary steam pipeline elbow and circumferential welded joint made of 13 HMF steel after approx. 200,000 h service under creep conditions for the operating parameters of further service.

3.6 Estimation of exhaustion degree caused by creep (step 6)

The exhaustion degree for materials working under creep conditions was defined in Chapter 3 of the study as the time of service until the time to rupture of this material under the temperature and stress operating conditions (t_e/t_r).

For the reviewed case, the residual life t_{re} (**Table 2**, column 9) was determined based on the developed relationship of $\log \sigma = f(t_{re})$ for the operating temperature of service T_e. It was determined for the level of operating stress σ_e of the test components as shown in **Figure 11**.

To determine the operating stress level using the formula

$$\sigma_e = \frac{p_e \cdot [D_z - g_{rz\,\min} \cdot (2 - z)]}{2 g_{rz\,\min} \cdot z}$$

where σ_e is the operating stress, $g_{act\ \min}$ is the actual minimum wall thickness of component, p_e is the operating pressure of component, D_{Out} is the nominal outside diameter, and z is the weakening coefficient, it is necessary to know the specific quantities in the formula.

Component	Operating parameters			Geometrical dimensions of component		Weakening coefficient z	Determined operating stress σ_r, MPa	Residual life t_{re}, h for σ_r	Exhaustion degree t_e/t_r
	Operating pressure p_e, MPa	Operating temperature T_e,	Service time t_e, hx10^{-3}	Outer diameter of component D_{out}, mm	Wall thickness nominal g_n actual $g_{act\ min}$				
1	2	3	4	5	6	7	8	9	**10**
elbow	13.8	540	200	324	$\frac{40}{36.8}$	1	51.5	175,000	**0.53**
Circumferential welded joint	13.8	540	200	324	$\frac{40}{40}$	1	46.9	165,000	**0.55**

Table 2.
Exhaustion extent t_e/t_r of the material of test primary steam pipeline components made of 13 HMF steel after 200,000 h service under creep conditions for the operating parameters of service.

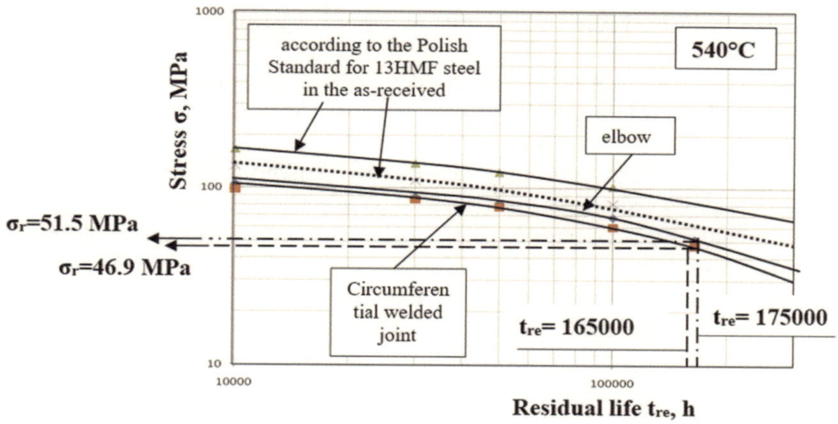

Figure 11.
Residual life t_{re} of the test primary steam pipeline components made of 13 HMF steel after 200,000 h service under creep conditions for the operating stress σ_e and operating temperature T_e.

Their designation and values for the test material of elbow and circumferential welded joint are summarised in **Table 2**, columns 2–7. The table also presents the operating stress σ_r values calculated for the test materials after service (column 8).

The life t_r is defined as the total time of the previous service t_e and residual life t_{re} for the parameters of the previous service (σ_e, T_e), that is, $t_r = t_e + t_{re}$. Since the exhaustion degree based on the above-mentioned definition can be written as t_e/t_r and $t_r = t_e + t_{re}$, then, in practice, $t_e/t_r = t_e/(t_e + t_{re})$. Using this relationship, the exhaustion degree (**Table 2**, column 10) was determined based on the residual life t_{re} determined and knowledge of the previous time of service.

4. Method for determination of the minimum component wall thickness necessary to transfer the actual service loads of the material after long-term service under creep conditions with known creep strength characteristics

The most important element necessary to determine the minimum required component wall thickness g_{oe}, which will be able to transfer the required service load (σ_{ep}, T_{ep}), is to have the characteristics of residual creep strength for the material of test component after service in the form of $\log \sigma = f(t_{re})$ for the operating temperature T_{ep}. This is the first step in the adopted procedure. Step 2 is to determine the residual creep strength $R_{Ze/Tep/tep}$ for the adopted temperature of further service T_e and the assumed time of further service t_e based on $\log \sigma = f(t_{re})$ at T_{ep} = const. The value of $R_{Ze/Tep/tep}$ obtained from the characteristics makes it possible to determine the acceptable stress k for the parameters of further service adopted in this way, which is step 3. To calculate the required minimum wall thickness g_{oe} for the adopted parameters of further service, the nominal outside diameter D_{Out} of the component, the operating pressure p_{ep}, and the construction weakening coefficient z should be defined, which is step 4. Step 5 is the determination of the required minimum wall thickness g_{oe} of the component for the adopted parameters of further service.

And the last, sixth, step is to compare the obtained value of the calculated required minimum wall thickness g_{oe} to the measured minimum actual thickness $g_{act\ min}$. If for the adopted time of further safe service t_{ep}:

- $g_{oe} < g_{act\ min}$, the component can be allowed for further service.
- $g_{oe} \geq g_{act\ min}$, the component cannot be allowed to continue the operation at the required parameters of further service for the adopted time t_{ep}.

If $g_{oe} \geq g_{act\ min}$ and the component cannot be allowed for further service, a shorter time of service should be adopted and the proposed procedure repeated. It may turn out that with a shorter time of further service t_{ep}, for the adopted parameters of further service, the condition of $g_{oe} < g_{act\ min}$ will be met. This will allow the test component to be operated for such a shorter time t_{ep}.

The proposed procedure in the form of an algorithm consisting of consecutive steps is shown in **Figure 12**. Its application is presented on the example of the primary steam pipeline elbow and circumferential welded joint made of 13 HMF steel after 200,000 h service.

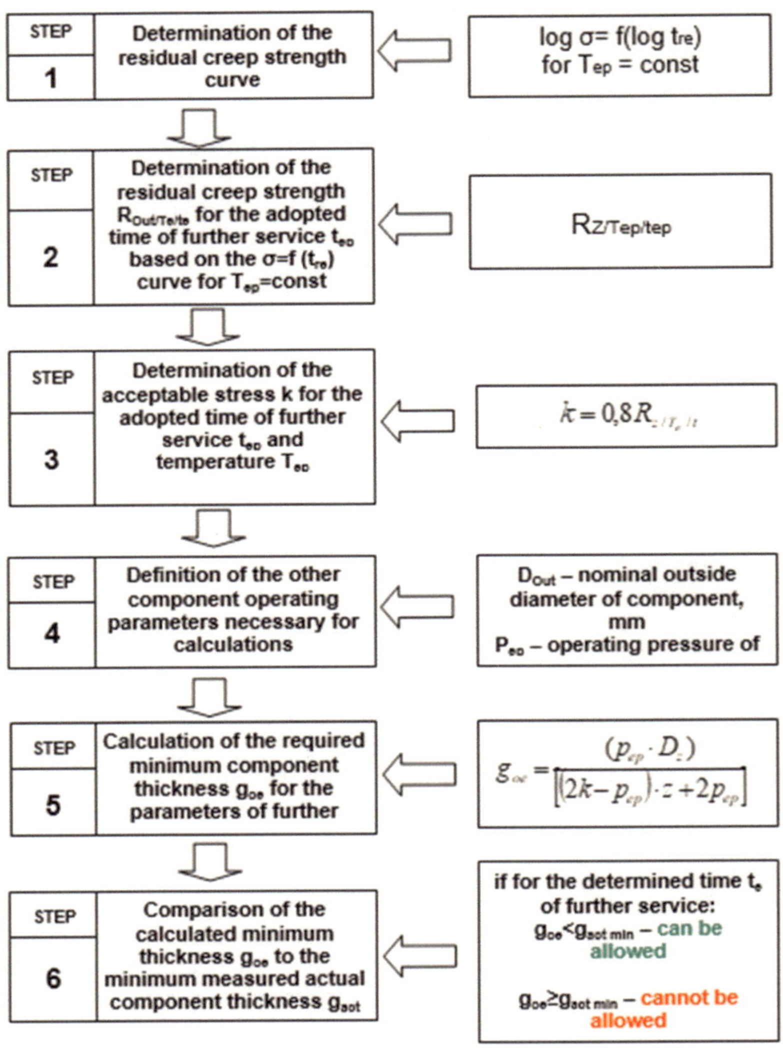

Figure 12.
Procedure for determination of the minimum wall thickness of a component after long-term service under creep conditions necessary to transfer the required service loads based on the residual life characteristics of the assessed material after service under creep conditions.

4.1 Characteristics of residual creep strength

The temporary creep strength curve for the material of elbow after 200,000 h service under creep conditions plotted based on results of the tests discussed in Chapter 2 is shown in **Figure 13** in the form of lg $\sigma = f(t_{re})$ for the temperature of further service $T_{ep} = 540°C$.

4.2 Determination of residual creep strength for the adopted time of further service

Based on the characteristics in point 2, the residual creep strength at 530 and 540°C for 10,000, 30,000, 50,000, and 100,000 h was determined for the materials of test primary steam pipeline components after 200,000 h service. The obtained results are presented in **Table 3**.

4.3 Determination of the acceptable stress

The value of acceptable stress k is equal to the value of acceptable lower scatter band −20% from the obtained mean value of residual creep strength for the adopted time t_{ep} and temperature T_{ep}. It is calculated from the following formula:

$$k = 0.8R_{Ze/Te/te}.$$

The obtained values of acceptable stress k for the adopted levels of temperature $T_{ep} = 530$, 540°C, and times $t_e = 10,000$, 30,000, 50,000, and 100,000 h of further service are provided in **Table 4**.

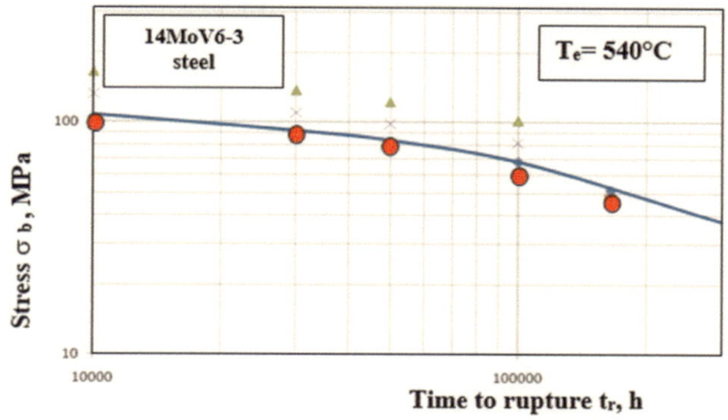

Figure 13.
Temporary residual creep strength of the material of test elbow and circumferential welded joint made of 14MoV6–3 steel after 200,000 h service at 530°C under creep conditions.

Component	Designation of properties	Residual creep strength R_{ze}, MPa at the test temperature $T_e = 540°C$
Elbow	$R_{ze/10,000}$	105
	$R_{ze/30,000}$	89
	$R_{ze/50,000}$	82
	$R_{ze/100,000}$	68

Table 3.
Residual creep strength for the adopted levels of temperature $T_{ep} = 530$, 540°C, and times $t_{ep} = 10,000$, 30,000, 50,000, and 100,000 h of further service of the materials of test elbow made of 14MoV6–3 steel after 200,000 h service under creep conditions.

Component	Adopted time of further service t_{ep}, h	Acceptable stress k, MPa for temperature of further service T_{ep} = 540°C
Elbow	10,000	84
	30,000	71.2
	50,000	65.6
	100,000	54.4

Table 4.
Determined values of acceptable stress k for the adopted levels of temperature T_{ep} and times t_e of further service for the material of test primary steam pipeline component made of 14MoV6–3 steel after 200,000 h service under creep conditions

4.4 Definition of parameters of further service

The parameters of further service are defined to be the same as the previous ones. To calculate the required minimum wall thickness, in addition to pressure p_{ep} and temperature T_{ep} of the component, it is also necessary to define its nominal outside diameter D_{Out}, nominal wall thickness g_n, design pressure p_o, operating pressure of further service p_e, and construction weakening coefficient z. The required values are summarised in **Table 5**.

4.5 Calculation of the required minimum component wall thickness for the parameters of further service

The calculation of the minimum component wall thickness for the parameters of further service g_{oe} was made based on the following formula:

$$g_{oe} = \frac{\left(p_{ep} \cdot D_z\right)}{\left[\left(2k - p_{ep}\right) \cdot z + 2p_{ep}\right]}$$

where g_{oe} is the minimum component wall thickness for the parameters of further service, p_{ep} is the operating pressure of further service, D_{Out} is the nominal outside diameter, k is the acceptable stress, and z is the weakening coefficient.

No.	Parameter	Unit	Value
1	Nominal outside diameter D_{Out}	mm	324
2	Nominal wall thickness g_n	mm	40
3	Minimum actual wall thickness $g_{act\ min}$	mm	36.8
4	Time of further service t_{ep}	h	30,000
			50,000
			100,000
5	Temperature of further service T_{ep}	°C	540
6	Design pressure p_o	MPa	14.2
7	Operating pressure of further service p_{ep}	MPa	13.8
8	Weakening coefficient z	—	1

Table 5.
Parameters of further service and geometrical dimensions of test pipeline elbow made of 14MoV6 steel after 200,000 h service.

Temperature of further service T_{ep} = 540°C			
Time of further service t_{ep}, h	**30,000**	**50,000**	**100,000**
Operating pressure of further service p_{ep}, MPa	**Required minimum wall thickness g_{oe}, mm**		
14.2	29.38	31.64	37.40
13.8	28.62	30.84	36.47

Table 6.
Required minimum wall thickness of primary steam pipeline elbow made of 14MoV6–3 steel for the adopted parameters of further service.

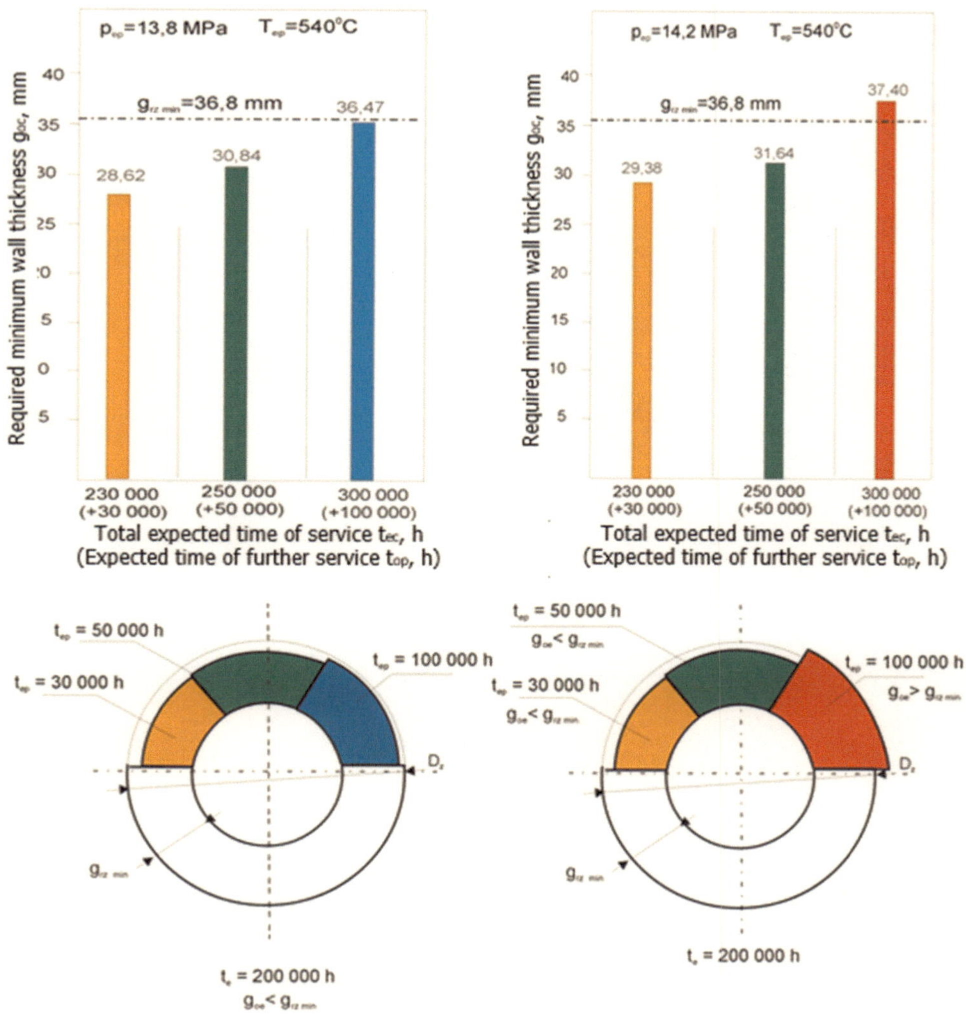

Figure 14.
Comparison of the required minimum pipeline elbow wall thickness g_{oe} for the adopted parameters of further service (p_{ep}, T_{ep}) to the minimum actual wall thickness $g_{act\ min}$ measured on the component for the adopted temperature of further service T_{ep} = 540°C.

The values obtained for pressure p_o and p_{ep} and for the adopted different times t_{ep} and temperature levels of further service T_{ep} for the material of test primary steam pipeline elbow made of 13HMF steel are summarised in **Table 6**.

4.6 Comparison of the calculated required minimum wall thickness to the actual minimum thickness based on measurements taken on the component

The obtained results of calculations of the required minimum wall thickness for the parameters of further service (**Table 6**) were compared to the actual minimum thickness based on measurements taken on the component (**Table 5**, item 3) as shown graphically in **Figure 14**.

These results meet the condition of $g_{oe} < g_{act\ min}$ except for the adopted temperature of further service T_{ep} = 540°C and the time of further service t_{ep} = 100,000 h for which this condition is not met and the component could not be allowed for further service of 100,000 h at 540°C for the other adopted operating parameters.

Author details

Janusz Dobrzański
Institute for Ferrous Metallurgy, Gliwice, Poland

*Address all correspondence to: jdobrzanski@imz.pl

References

[1] Sroka M, Zieliński A, Mikuła J. The service life of the repair welded joint of Cr-Mo/Cr-Mo-V. Archives of Metallurgy and Materials. 2016;**61**: 969-974

[2] Zieliński A, Golański G. The influence of repair welded joint on the life of steam pipeline made of Cr-Mo steel serviced beyond the calculated working time. Archives of Metallurgy and Materials. 2015;**60**:1045-1049

[3] Zieliński A, Dobrzański J, Purzyńska H, Golański G. Properties, structure and creep resistance of austenitic steel super 304H. Materials Testing. 2015;**57**: 859-865

[4] Golański G, Jasek J, Zieliński A, Kolan C, Urzynicok M, Wieczorek P. Quantitative analysis of stability of 9% Cr steel microstructure after long-term ageing. Archives of Metallurgy and Materials. 2017;**62**:273-281

[5] Sroka M, Zieliński A, Dziuba-Kałuża M, Kremzer M, Macek M, Jasiński A. Assessment of the residual life of steam pipeline material beyond the computational working time. Meta. 2017;7:3

[6] Dziuba-Kałuża M, Zieliński A, Dobrzański J, Sroka M, Urbańczyk P. Residual life of boiler pressure parts made of the 13CrMo4-5 steel after long-term operation in a creep conditions. Archives of Metallurgy and Materials. 2018;**63**(2):889-897

[7] Purzyńska H, Golański G, Zieliński A, Dobrzański J, Sroka M. Precipitation study in Ti-stabilised austenitic stainless steel after 207,000 h of service. Materials at High Temperatures. 2019. DOI: 10.1080/09603409.2018.1546919

[8] Zieliński A, Sroka M, Dudziak T. Microstructure and mechanical properties of Inconel 740H after long-term service. Materials. 2018;**11**:2130. DOI: 10.3390/ma11112130

[9] Zieliński A, Dobrzański J, Purzyńska H, Sikora R, Dziuba-Kałuża M, Kania Z. Evaluation of creep strength of heterogeneous welded joint in HR6W alloy and Sanicro 25 steel. Archives of Metallurgy and Materials. 2017;**62**(4): 2057-2064

[10] Sroka M, Zielinski A, Hernas A, Kania Z, Rozmus R, Tanski T, et al. The effect of long-term impact of elevated temperature on changes in the microstructure of inconel 740H alloy. Metalurgija. 2017;**56**(3–4):333-336

[11] Sroka M, Nabialek M, Szota M, Zielinski A. The influence of the temperature and ageing time on the NiCr23Co12Mo alloy microstructure. Revista de Chimie. 2017;**68**(4):737-741

[12] Golanski G, Zielinska-Lipiec A, Zielinski A, Sroka M. Effect of long-term service on microstructure and mechanical properties of martensitic 9% Cr steel. Journal of Materials Engineering and Performance. 2017; **26**(3):1101-1107

[13] Zieliński A, Golański G, Sroka M. Influence of long-term ageing on the microstructure and mechanical properties of T24 steel. Materials Science & Engineering, A: Structural Materials: Properties, Microstructure and Processing. 2017;**682**:664-672

[14] Zieliński A, Golański G, Sroka M. Comparing the methods in determining residual life on the basis of creep tests of low-alloy Cr-Mo-V cast steels operated beyond the design service life. International Journal of Pressure Vessels and Piping. 2017;**152**:1-6. DOI: 10.1016/ j.ijpvp.2017.03.002

[15] Dobrzański J, Hernas A. Correlation between phase composition and life-

time of 1Cr-0.5Mo steels during long term service at elevated temperature. Journal of Materials Processing Technology. 1995;**53**:101-108

[16] Dobrzański J, Hernas A. Relationship between microstructure and remanent life-time of low alloy Cr-Mo steels. In: 6th International Conference on Creep, Fatigue and Creep-Fatigue Interaction. London; 1996. p. 451

[17] Dobrzański J. The classification method and the technical condition evaluation of the critical elements' material of power boilers in creep service made from the 12Cr-1Mo-V. Journal of Materials Processing Technology. 2005;**164-165**:785-794

[18] Dobrzański J, Hernas A, Pasternak J, Zieliński A. Microstructure and mechanical properties characteristics of welded joints made of creep-resistant steel with 12% Cr, V, W and Co additions. In: Proceedings of the 5th International Conference on Advances in Materials Technology for Fossil Power Plants, 3–5 October 2007. Marco Island, Florida: ASM International; 2008. pp. 303-319

[19] Zieliński A, Dobrzański J, Renowicz D, Hernas A. The estimation of residual life of low-alloy cast steel Cr-Mo-V type after long-term creep service. In: Proceedings of the 5th International Conference on Advances in Materials Technology for Fossil Power Plants, 3–5 October 2007. Marco Island, Florida: ASM International; 2008. pp. 616-626

[20] Neubauer B. Creep damage evolution in power plants. In: Proc. Symp. Creep and Fracture of Eng. Mat. and Structures. Swansea; 1984. p. 2

[21] Dobrzański J. Internal damage processes in low alloy chromium-molybdenum steels during high-temperature creep service. Journal of Materials Processing Technology. 2004; **157-158**:297-303

[22] Dobrzański J, Sroka M, Zieliński A. Methodology of classification of internal damage the steels during creep service. Journal of Achievements in Materials and Manufacturing Engineering. 2006; **18**:263-266

[23] Dobrzański J, Sroka M. Computer aided classification of internal damages the chromium-molybdenum steels during creep service. Journal of Achievements in Materials and Manufacturing Engineering. 2007; **24**(3):143-146

[24] Zieliński A, Dobrzański J, Krztoń H. Structural changes in low alloy cast steel Cr-Mo-V after long time creep service. Journal of Achievements in Materials and Manufacturing Engineering. 2007; **25**(1):33-36

[25] Dobrzański J, Krztoń H, Zieliński A. Development of the precipitation processes in low-alloy Cr-Mo type steel for evolution of the material state after exceeding the assessed lifetime. Journal of Achievements in Materials and Manufacturing Engineering. 2007; **23**(2):19-22

[26] Dobrzański J, Zieliński A, Krztoń H. Mechanical properties and structure of the Cr-Mo-V low-alloyed steel after long-term service in creep condition. Journal of Achievements in Materials and Manufacturing Engineering. 2007; **23**(1):39-42

[27] Zieliński A. The effect of long-term service under creep conditions on mechanical properties of Cr-Mo-V cast steel. In: 8th International Scientific Conference-New Technologies and Achievements in Metallurgy and Materials Engineering. Vol. II. Czestochowa: Publishing House of the Czestochowa University of Technology; 2007. pp. 717-721

[28] Hernas A, Dobrzański J. Life Time and Destruction of Steam Boilers and Turbines. Gliwice: The Silesian University of Technology Publishing House; 2003

Chapter 6

Physical Background and Simulation of Creep in Steels

Stan T. Mandziej

Abstract

The simulative accelerated creep test (ACT) was developed as a response to an overall need of gaining in a short time useful physical data for determining long-term behavior of materials exposed to operation under stress at elevated temperatures in power generation and chemical processing industries. Additionally, the recently frequent power plant shutdowns due to adding solar/wind power to the net, call for creep-fatigue data, which standard creep tests cannot provide. In response to these needs, a thermal-mechanical fatigue procedure-ACT-was designed, taking into account physical phenomena causing microstructure transformation during creep, in particular generation of dislocation substructures, their role in nucleation of voids and cracks, intensification of carbide precipitation, and decay of mechanical properties during long-time exposure to elevated temperatures. The actual ACT procedure generates adequate data for calculating true lifetime of the tested creep resisting material for a nominal stress.

Keywords: creep, physical simulation, dislocation substructures, carbide precipitation

1. Introduction

Conventional "constant-load" creep tests of new steels and welds for thermal power generation and chemical processing are very long lasting, thus delaying the application of newly developed steels for years. Even more serious problem appears with creep tests on heat-affected zones in welded joints of these steels, as welding procedure seriously influences the creep properties and the HAZs are very often considered as weakest links of the joints. While most of the design of components and estimation of power plants lifetime is based on long-term creep data, which are generally available for the plate and pipe creep resisting steels, in case of the welded joints on these components, such data are not always easily found, for example, the repair welds are very often not available, especially when repairs were performed outside. To address all these issues, a simulative test was developed allowing to obtain creep life data in less than 24 h. It has been implemented on the Gleeble physical simulator [1]. It is based on detailed observations of microstructural phenomena occurring in martensitic-ferritic steels and welds during exposure to creep. The accelerated creep test is a low-cycle thermal-mechanical fatigue procedure designed to transform the creep resisting steels and welds in the same manner like true creep, only occurring in much shorter time.

2. Physical simulation of creep

Considering creep as a plastic deformation appearing at elevated temperature, with small strains and very small strain rates [2], next to the short duration demanded from the ACT are the following conditions:

1. The basic temperature and applied strains in the ACT must prevent odd transformations like secondary dissolution of carbides or intensive formation of nonequilibrium phases, as well as strain-induced precipitation.

2. The final deformation at fracture, in particular crosswise strain, must be like at real creep-just a few pct in total.

3. The depletion of steel matrix in alloying elements must be achieved similar to that of crept steels, and the carbide phases or other precipitates at onset of cracks must not be different.

4. As creep in the matrix of ferritic-martensitic steels is the phenomenon in which thousands of grains or hundred of thousands of subgrains take part, the portions of samples for ACT that undergo transformation must have adequate "bulk" size, so the first creep voids and cracks in the test nucleate predominantly in most vulnerable sites of microstructure.

Then the simulative test must be in agreement with the following observations:

The creep voids nucleate preferentially at grain boundaries of recrystallized grains or recovered subgrains, in which oriented slip is observed (**Figure 1**), while inside several creep voids, escape of slip lines can be seen (**Figure 2**). Such voids very seldom nucleate at any larger precipitates.

In the substructure of the crept steels, *a*/2<111> screw dislocations dominate, and these dislocations when not pinned to precipitates are highly mobile. In thin foil of thickness about 260 nm, relatively medium density of such dislocations could be seen (**Figure 3**), while in thicker foil of about 380 nm, their density is much higher, especially when interacting with precipitates and on intersections of two families of these screw dislocations, the *a*<001> edge dislocations could be identified (**Figure 4**). These last dislocations are mobile with Burgers vectors in {110} planes [3].

When the sample taken from a crept material containing internal "clean" creep voids not exposed to atmosphere is fractured below its ductile-to-brittle transition

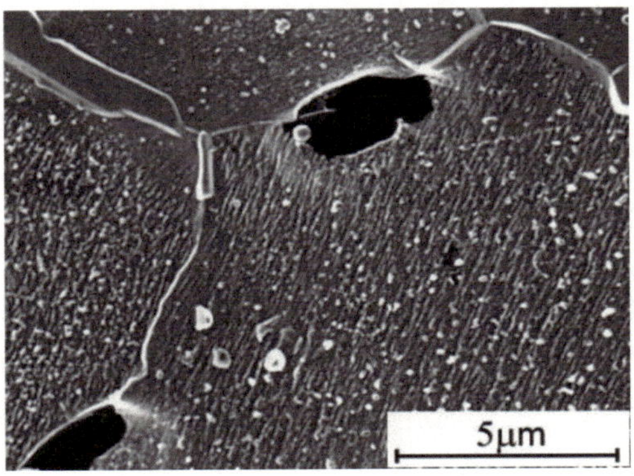

Figure 1.
½ CrMoV steel crept for 22.5 years/16.6 MPa at 568°C; $FeCl_3$ etched for slip lines, SEM SE-image.

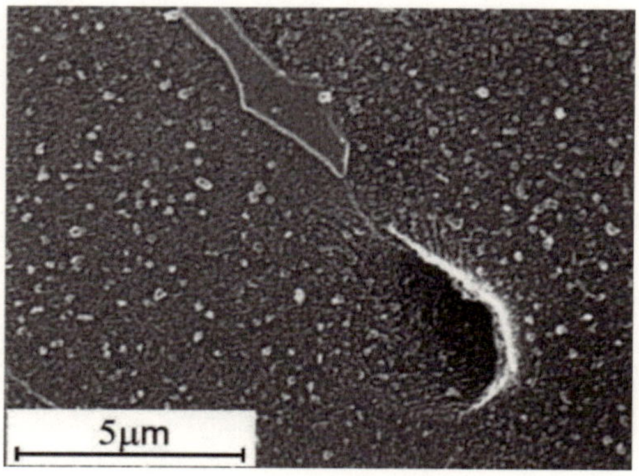

Figure 2.
½ CrMoV steel crept for 22.5 years/16.6 MPa at 568°C; $FeCl_3$ etched, SEM SE-image; slip lines visible inside creep void.

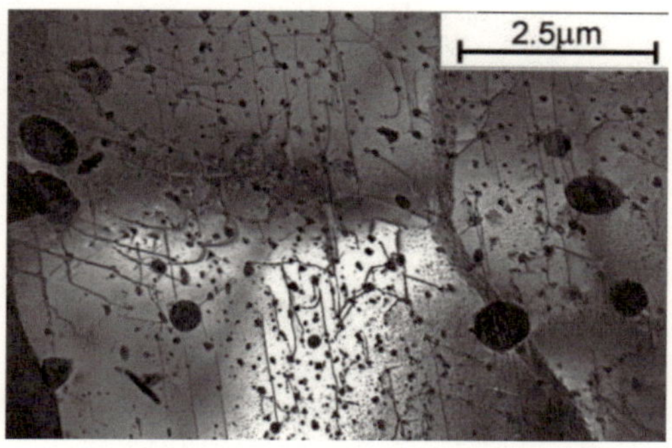

Figure 3.
½ CrMoV steel crept for 20.5 years at 568°C; precipitates and a/2 <111> dislocations in ferrite; TEM, thin foil ~260 nm thick.

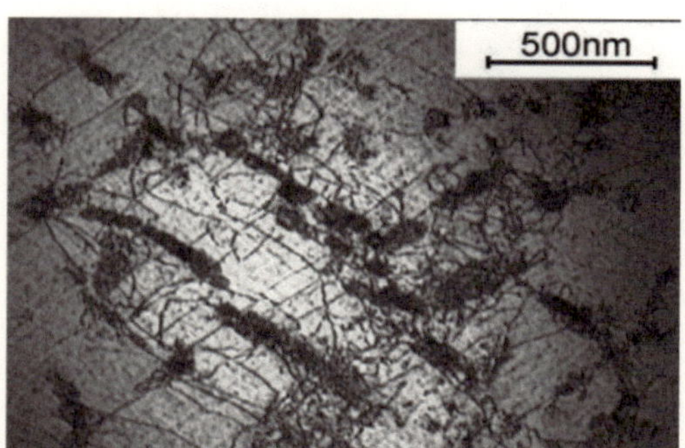

Figure 4.
½ CrMoV steel crept for 20.5 years at 568°C; a/2 <111> dislocations interacting with precipitates; TEM, thin foil ~380 nm thick.

temperature, the voids open up, and on their walls, a characteristic pattern of criss-crossing perpendicular lines can be seen (**Figure 5**) [4]. This evidence confirms the micromechanism of creep microcracks nucleation by piling up mobile dislocations in slip planes at grain boundaries [5] and, after identifying the dislocations and their glide planes, tells that these slip planes are two perpendicular of {110} type. The schematic two-dimensional drawing in **Figure 6** is a model illustrating this situation.

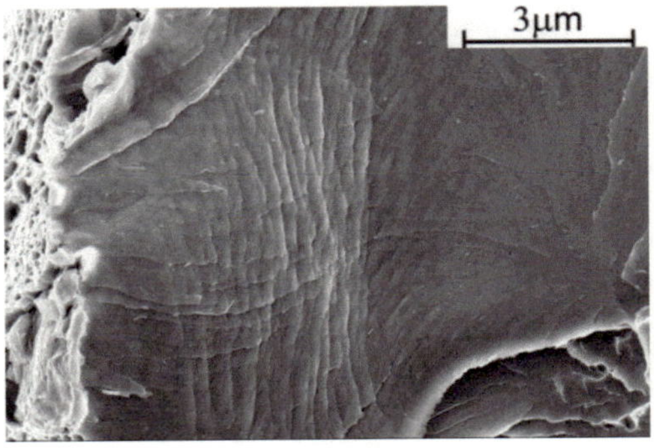

Figure 5.
Creep crack surface with numerous criss-crossing slip lines, in ½ CrMoV steel exploited for 22.5 years at 568°C; SEM image, no etching.

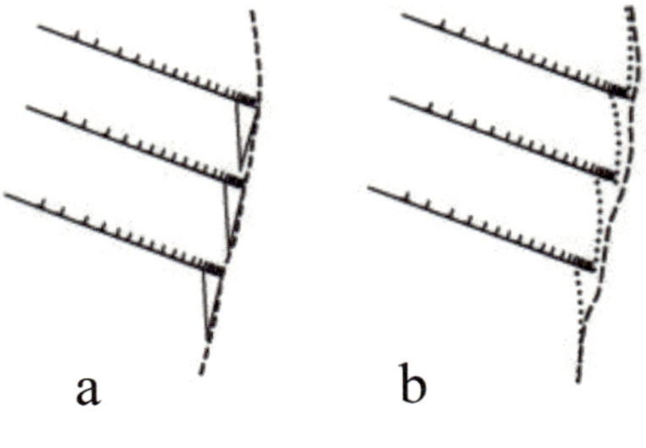

Figure 6.
Schematic drawing of voids (a) and cracks nucleation by the voids coalescence (b) caused by dislocations piled up at grain boundary.

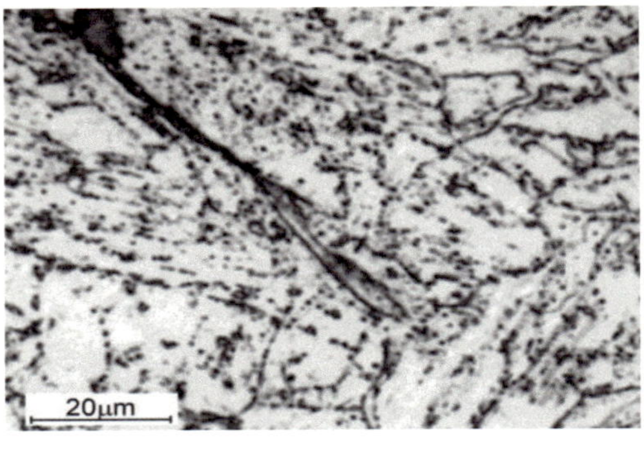

Figure 7.
Microcrack developed near HAZ in P91 steel after service for ~3 years at 568°C; light microscope image, $FeCl_3$ etched.

The *a*<001> edge dislocations in steel ferrite matrix are known from earlier work as effectively transporting interstitial elements, thus enhancing precipitation rate. During "in-situ" heating experiments in TEM on HSLA steel specimens containing preformed configurations of such dislocations, ultrafine precipitates in fractions of seconds appeared in sites of these dislocations escape and annihilation [6]. In prematurely failed P91 steel higher than average density of fine precipitates appears near to cracks (**Figure 7**). And the generation of substructures containing large amount of such *a*<*001*> edge dislocations was observed during deformation of dual-phase steels [7], and also they appeared in this prematurely failed P91 steel (**Figure 8**).

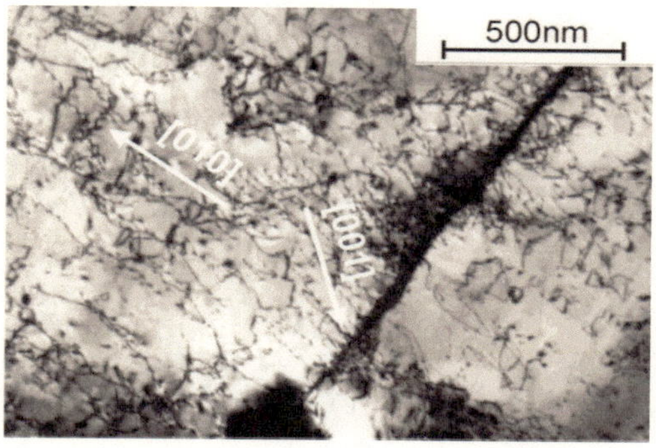

Figure 8.
Substructure of ferrite containing a<001> *edge dislocations in P91 steel after service for ~3 years at 568°C; TEM, thin foil.*

3. The simulative accelerated creep test

Taking into account the mentioned phenomena associated with creep and implementing the factors speeding up microstructure transformation, the simulative accelerated creep test was designed. It relies on the response of tested material, and for intensifying transformation of microstructure, various programmed testing procedures are used. The ACT was implemented on Gleeble physical simulator using samples with central gauge portion of 12 mm length × 10 mm diameter, mounted like in **Figure 9**. When the span between the "cold" mounting copper jaws is 32–35 mm, the uniformly-heated zone of about 10 mm length is created in the middle of the sample, due to direct resistance heating of the Gleeble balanced by controlled heat flow toward the "cold" Cu jaws. Sizes of the ACT samples are given in **Figure 10**. For testing base steel without weld, plain samples of the same geometry shall be used.

The ACT samples mounted in the Gleeble's "pocket jaws" assembly, like in **Figure 9**, are subjected to programmed cycles of low-cycle thermal-mechanical fatigue, run till failure or till predetermined stress or strain. Loading of the samples is executed by displacement (stroke) control and response of the sample is recorded as force and calculated engineering stress. The ACT procedure uses small elastoplastic tensile and compressive strains applied to central portion of rod-like samples (**Figure 10**) subjected to multiple thermal cycles at temperatures

Figure 9.
Mounting of the ACT sample in "cold" Cu-jaws set of Gleeble, with TC percussion welded in the middle of gauge length.

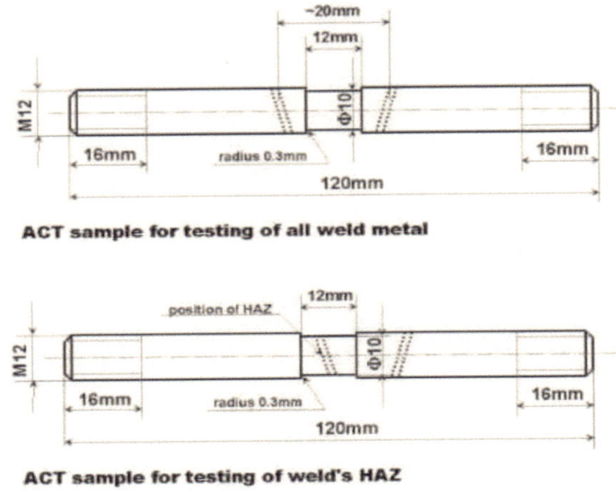

Figure 10.
Schematic drawing of cross-weld samples used for ACT, for testing all-weld-metal and for weld's HAZ.

characteristic of creep. In homogeneous samples, like plain steel bars or all weld metals, often before the crack appearance on the surface, internal cracks are formed perpendicular to the sample's axis. In microstructurally inhomogeneous samples, the cracks follow the "weakest links" like the HAZs of the welds. The ACT speeds up microstructural changes by accumulating elastoplastic tensile and compressive strains in the central portion of the sample during thermal cycling at temperatures characteristic of creep. The compressive strains result in a decrease of yield strength by generation of mobile dislocations, that is, the Bauschinger effect, and this accelerates creep [8]. Then, in the second part of the loading cycle, that is, changing the load to tension, a relaxation time under constant load is applied, during which the sample extends with strain rates equivalent to 10^{-5} to 10^{-8} per second, depending on its stiffness (pseudoelasticity modulus E*) and yield strength.

The shape of the sample's gauge portion after the ACT interrupted before fracturing is given in **Figure 11**. When the heat flow is symmetric to both cold clamping jaws and the tested material is homogeneous, centrally located slight necking forms on the sample often coinciding with an internally nucleated crack (**Figures 11a** and **12a**). **Figure 12b** shows on cross-section of this sample the size of transient zone (TZ) between the grip portion and the uniformly-heated central zone (UHZ). In the case of testing cross-weld sample with weld HAZ, the crack

Figure 11.
Shape of the cross-weld, all-weld-metal sample after the ACT terminated before entire fracturing (a), and of weld's HAZ with the crack following the HAZ's direction (b).

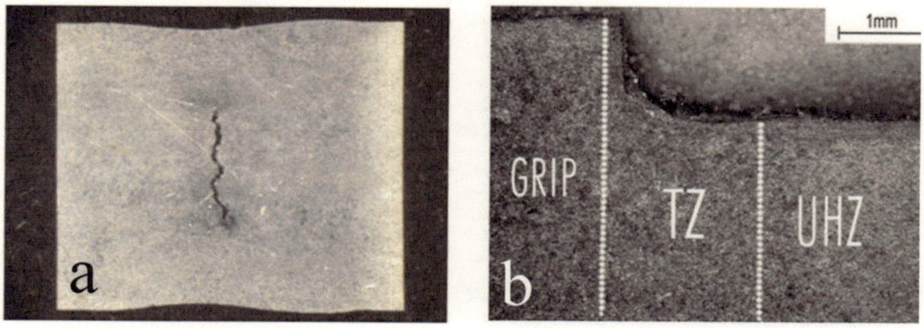

Figure 12.
An internal crack formed in the sample perpendicular to loading axis (a) and size of transient zone between the mounting portion and the gauge portion on the ACT sample (b).

appearing on the surface is often inclined to the longitudinal axis of the sample, following the HAZ (**Figure 11b**).

4. Results of ACT

A characteristic of creep is the decrease of steel hardness with the progress of time. Therefore, any creep test that results in the increase of hardness of sample must be considered as nonreliable. **Table 1** contains examples of initial values for nonexploited materials, for creep exploited materials and for materials subjected to ACT. In the first instance, it shows that during real creep the hardness is most dependent on the time of exposure as well as on the applied stress. Then, the important factor is the initial hardness. Finally, it shows that the drop of hardness after just a few thousand seconds in the ACT follows the tendency of multiyear real creep.

Usually, the tests are run till failure/fracturing, but they can also be stopped anytime and specimens for metallographic, fractographic, and microanalytical investigations can be taken before fracturing. Here, an example is given of the 9Cr-1.5Mo-1.3Co-V-Nb-N-B (FB2) steel, which in as-delivered state has a high-tempered coarse martensitic microstructure (**Figure 13**), while in its as-welded and postweld heat-treated HAZ, such microstructure changes into fine grained, nonacicular with distinguished grains of ferrite. In this pseudo dual-phase microstructure examined by ACT on crosswise samples, microcracks were found when the ACT was interrupted at about 90% of its duration. Such cracks, as shown in **Figure 14**, were nucleating in intercritically-reheated fine grain portion of heat-affected zone, and they often initiated at specific locations of the HAZ, where two subsequent heat cycles of multipass welding overlapped, giving dual-phase microstructure.

As the tests for different materials are run at different temperatures and the response of material results in various stresses equivalent to the yield strength at elevated temperature of the test, to compare the ACT results, the duration of the test and its temperature can be included in the following parameter:

$$P_{ACT} = (7 + \log t) \times T/100 \tag{1}$$

where t = time of test in [ks] and T = temperature in [K].

The ACT result can be inserted onto a graph, having P_{ACT} on horizontal axis and average tensile relaxation stress R_S on vertical axis. Alternatively, a creep strength

No.	Material—sample/exposure or test	Hardness HV0.1/(HV30)	
		Initial or (standard)	After exposure or testing
1	1/2CrMoV pipe-exp. 22.5 years at 568°C and 16.6 MPa	(180)	(131)
2	1/2CrMoV pipe-exp. 20.5 years at 568°C and 3.8 MPa	(180)	158
3	P91 antler-exp. ~9 years at 600°C and 16.5 MPa	(250)	229
4	P91 weld-exp. ~9 years at 600°C and 16.5 MPa	(265)	208
5	P91 rolled pipe-exp. ~3 years at 568°C	(250)	231
6	P91 weld with PWHT-exp. ~3 years at 568°C	(265)	258
7	P91 forged bottle-exp. ~3 years at 568°C	(260)	243
8	12Cr1MoV-exp. 16 years, 550°C, 15 MPa + ACT 600° C, 19.5 ks	188	164
9	15Cr1Mo1V-exp. 18 years, 550°C, 15 MPa + ACT 600° C, 30.4 ks	176	157
10	P22 rolled pipe + PWHT/ACT 600°C, 9.5 ks	234	188
11	P22 weld metal + PWHT/ACT 600°C, 11.3 ks	238	202
12	P22 rolled pipe HAZ + PWHT/ACT 600°C, 6.4 ks	227	196
13	P91-std. weld with PWHT/ACT 600°C, 28.3 ks	287	230
14	P91-std. weld with PWHT/ACT 600°C, 82.6 ks	285	223
15	P91-std. weld with PWHT/ACT 625°C, 37.6 ks	285	217
16	P91-temper-bead weld, no PWHT/ACT 620°C, 51.3 ks	298	193
17	P92-std. weld with PWHT/ACT 625°C, 21.3 ks	294	240
18	P92-rolled pipe HAZ/ACT 625°C, 22.1 ks	268	228

Table 1.
Examples of hardness for steels subjected to creep and/or ACTested.

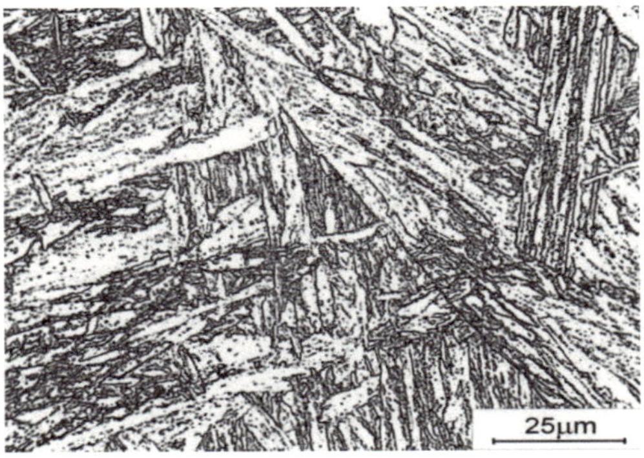

Figure 13.
High-tempered martensite of forged 9Cr-1.5Mo-1.3Co-V-Nb-N-B (FB2) steel in as-delivered state, $FeCl_3$ etched.

factor in ACT can be given as: $F_{ACT} = P_{ACT} \times R_S/100$ and used for comparing materials tested at the same conditions.

During the ACT procedure, the data of stress, strain, strain rate, and temperature, as well as dilatometric information, can be recorded, out of which strain-time and stress-time graphs are produced as shown in **Figures 15** and **16**.

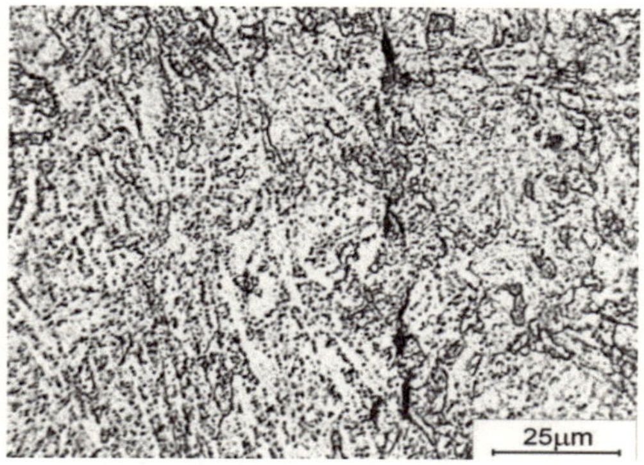

Figure 14.
Microcracks in IC-HAZ of FB2 steel perpendicular to load axis of ACT sample, developed at ~85% of test duration at 625°C.

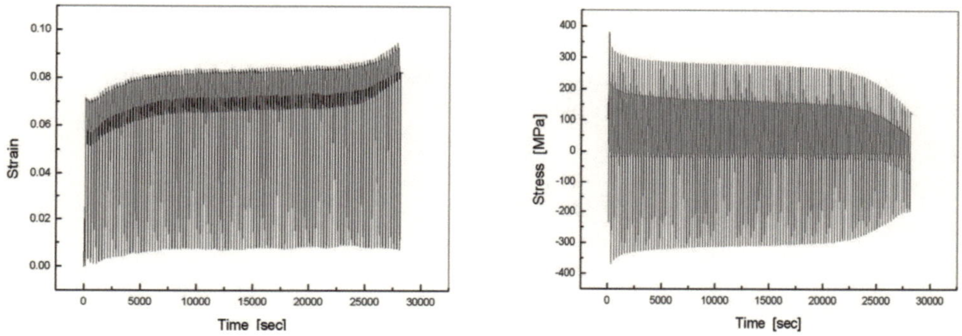

Figure 15.
Typical strain-time and stress-time graphs from ACT on a "soft" alloy, for example, postweld heat-treated P91 weld metal.

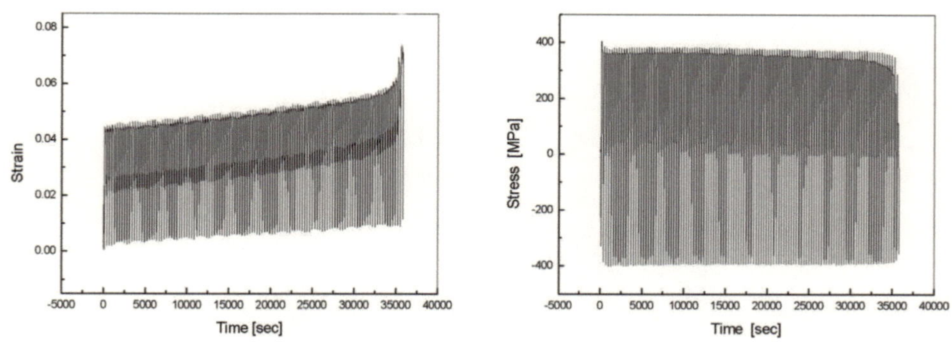

Figure 16.
Typical strain-time and stress-time graphs from ACT on a "hard" alloy, for example, as-deposited (without PWHT) P91 weld metal.

The results of ACT are usually plotted in the form of strain-time and stress-time graphs, and data from these graphs are used to calculate creep strength and true creep life. The "zero-stress" line on strain-time graph from ACT well resembles the true creep graph from conventional creep tests.

Examples of results obtained at different ACT temperatures for various P91 weld metals in as-welded state (AW) and postweld heat treated (HT) are given in **Table 2**.

The result of P91-HT2 is complimentary to sample #16 from **Table 1**, to which after temper bead welding the PWHT was applied for 2 h at 720°C in furnace, and then the ACT was run. Longer durations of some tests (P91-HT2 and P91-HT5) resulted from smaller loading displacements applied in such lower intensity ACT program. The tests are usually run at prevacuum of about 10^{-3} Tr maintained in the testing chamber of physical simulator; however, when clean fracture surfaces are expected after the ACT, high vacuum of better than 10^{-5} Tr may be applied. In these experiments on P91 steel, no difference in test duration due to vacuum level was observed. Visible difference in test duration was noted when water mist became introduced into the testing chamber; the result of P91-HT5m in **Table 2** shows shorter test duration, lower tensile relaxation stress, and in consequence lower creep strength factor. On cylindrical surface of this sample's gauge portion, numerous fine cracks appeared (**Figure 17**), which for the sample tested in high vacuum could not be seen up to the macrocrack appearance on the surface (**Figure 18**).

To verify the results of ACT by proving that the test adequately simulates situations appearing in real creep, metallographic investigations were carried out. **Figure 19** shows dislocation substructure of P91 weld metal at half-life of the ACT. In recovered postmartensitic subgrains criss-crossing, $a/2$ <111> screw dislocations appear in vast amount and form planar arrays.

Material and state	ACT temp. [°C]	ACT time to fracture [ks]	Tensile relaxation stress R_S [MPa]	Creep strength factor F_{ACT} [MPa]
P91-AW1	625	22.0	305	228
P91-HT1	625	37.6	273	210
P91-AW2	620	38.0	395	303
P91-HT2	620	51.3	177	138
P91-HT3	600	28.3	325	240
P91-HT4	600	26.3	318	233
P91-HT5	600	82.6	336	262
P91-HT5m	600	57.2	266	203

Table 2.
Examples of ACT results.

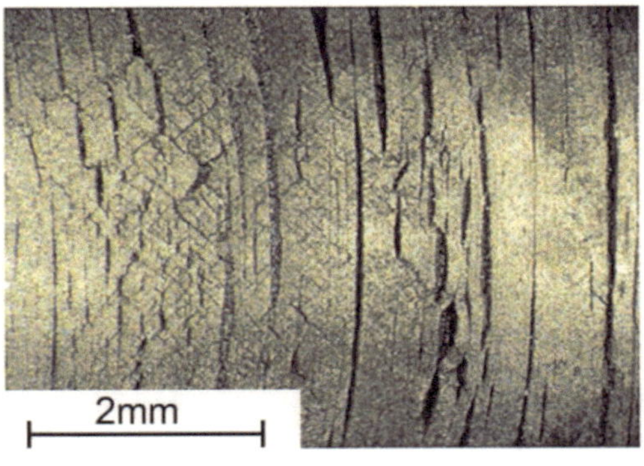

Figure 17.
Cylindrical surface of gauge portion on P91 ACT sample after test at 600°C with the presence of water mist in working chamber.

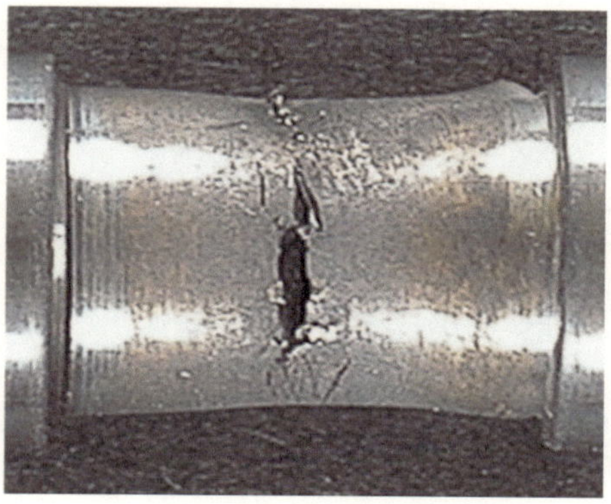

Figure 18.
Macrocrack on ACT sample tested at 600°C with high vacuum in simulator's working chamber.

Figure 19.
Subgrains with recovered dislocation configurations at half-life in ACT, in P91 weld metal sample tested at 600°C; TEM, thin foil.

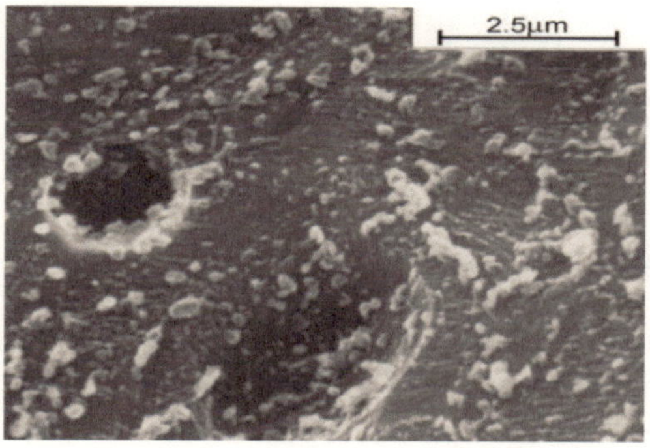

Figure 20.
Fracture surface of ACT sample of P91 steel showing high density of carbides, electro-polished and $FeCl_3$ etched, SEM image.

In **Figure 20**, the fracture surface of ACT sample is shown, after the removal of surface oxides by electropolishing and then etching. On this surface, the traces of slip lines can be recognized, thus confirming the fracture nucleation by slip and piling up dislocations in planar arrays.

The final verification has to be done on TEM images of the tested steel or weld metal before and after the ACT, like on the examples presented here in

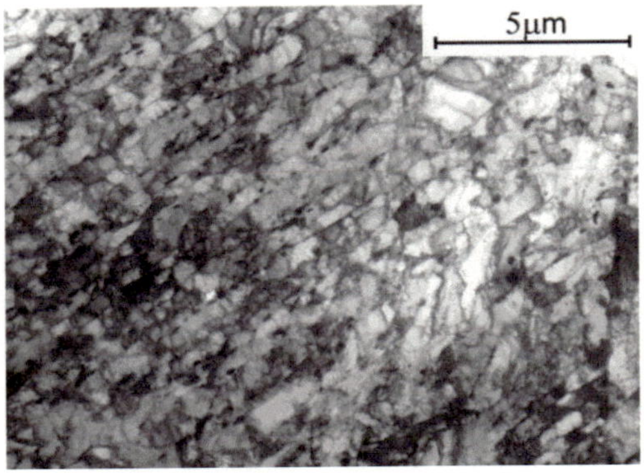

Figure 21.
Arrays of subgrains with carbides in the initial microstructure of P91 weld metal before the ACT; TEM, thin foil.

Figure 22.
Well-recrystallized ferrite grains with arrays of carbides, in P91 weld metal sample after completed ACT at 600°C; TEM, thin foil.

Figures 21 and **22**. These figures show the microstructure transformation in P91 weld metal from tempered lath martensite before the ACT to fully recrystallized after the ACT equiaxed ferrite grains with coagulated and spheroidized carbides, the last aligned in directions following prior martensite laths.

5. Combination of weld's HAZ simulation with ACT

The ACT procedure appeared useful to study other high strength steels for elevated temperature applications different than thermal power generation and chemical processing.

Table 3 contains results of ACT on nuclear reactor pressure vessel steel grade 15Ch2MFA (Russian), of which creep resistance is not the most important property; however, the strength of its HAZ after welding and PWHT is crucial. Physically-simulated HAZs on this steel were earlier studied and reported, as regards their microstructures and CVN impact strength [9]. The initial microstructure of 15Ch2MFA steel was high-tempered martensitic (**Figure 23**), and after the single-cycle 900°C physical simulation of HAZ, it changed into a ferritic-martensitic mixture in the middle of the simulated HAZ (**Figure 24**), with well-distinguished equiaxed fine bright ferrite grains and darker-etching areas, details of which could not be revealed by light microscopy.

	Material	ACT temperature [°C]	Time to fracture [ks]	Average tensile stress [MPa]	Creep life for 100 MPa [h]
1	Base 15Ch2MF steel	625	10.18	203	4183
2	15Ch2MF, HAZ 1-cycle simulated, no PWHT	625	8.81	235	2810
3	15Ch2MF, HAZ 1-cycle simulated + PWHT	625	12.37	187	5547
4	15Ch2MF, HAZ 2-cycle simulated + PWHT	625	8.12	264	9765
5	15Ch2MF, HAZ* 2-cycle simulated + PWHT	625	10.66	183	5873

Table 3.
Creep life of simulated HAZs in 15Ch2MFA steel.

Figure 23.
Initial high-tempered martensitic microstructure of 15Ch2MFA steel.

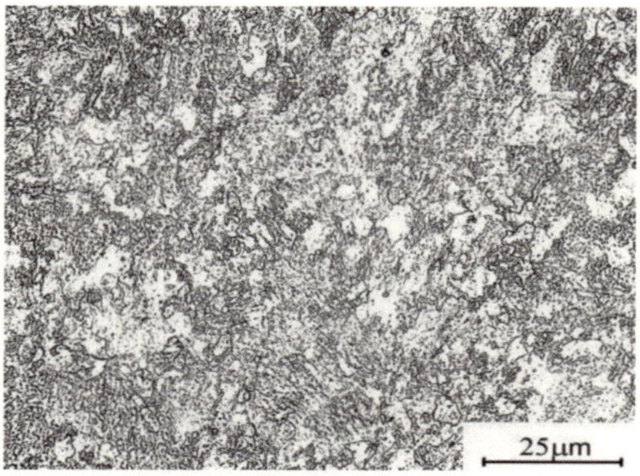

Figure 24.
Simulated intercritically-reheated microstructure of 15Ch2MFA steel HAZ.

More informative as regards components of this intercritical HAZ are TEM images from thin foil specimens, which reveal small compact martensitic islands with nondissolved carbides in the HAZ region representative to temperatures just above the A1 (**Figure 25**) and larger fields of homogeneous acicular martensite surrounded by ferrite grains (**Figure 26**) in the region closer to the 900°C peak

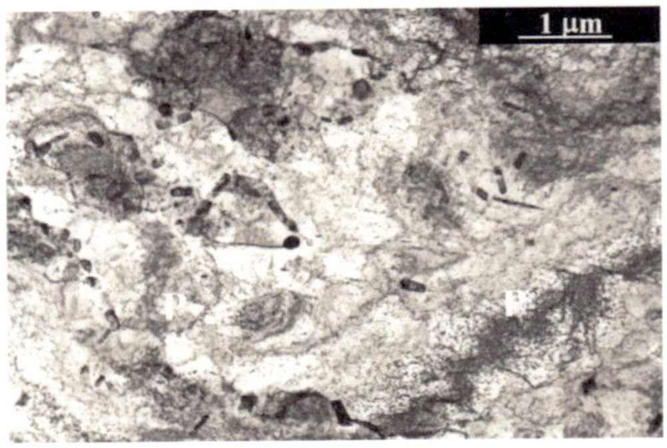

Figure 25.
Ferritic-martensitic microstructure with retained carbides in simulated HAZ region close to A1 (760°C) temperature.

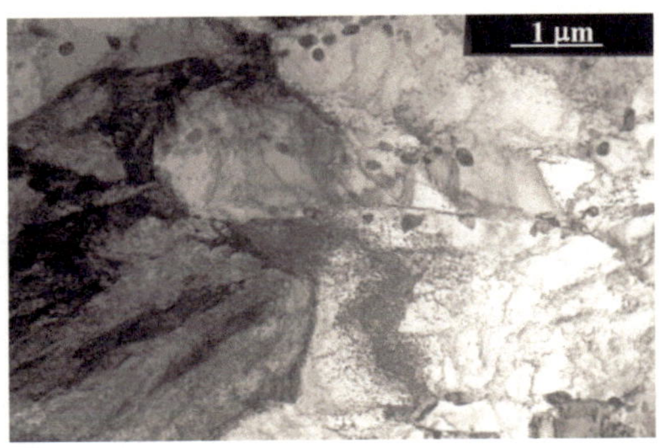

Figure 26.
Martensite grain and recrystallized ferrite, in simulated HAZ region close to peak (900°C) temperature.

temperature of the simulation cycle. In both cases, the ferrite interacting with martensite contained very high dislocation density. More details of this microstructure in as-simulated state and after conventional tempering were given in previous IIW-Document [9].

Creep life of the weld's simulated HAZ on this steel was tested using the simulated welding thermal cycle(s) followed immediately by ACT, applying setups as shown in **Figures 27** and **28**. Various simulations of the welding thermal cycles were carried out and then, by manipulation of the stress intensity in ACT, the

Figure 27.
The ACT sample in Gleeble chamber during simulation of HAZ's 1st cycle.

Figure 28.
The same sample in Gleeble chamber during simulation of HAZ's 2nd cycle.

strain-induced precipitation was achieved revealing potential of this steel to strengthen in the HAZ. However, not only the stress intensity but also the locations of post dual-phase microstructure generated by subsequent weld thermal cycles, that is, the second cycle of 920°C peak after the first cycle of 1220°C peak, contributed to the exceptionally high creep life of the HAZ in the case of sample #4 (result in row 4 of **Table 3**). In classical welding HAZ simulation experiments, the thermal cycles are executed having their peak temperatures in the same locations on the fixed sample (see **Figures 27** and **28**), and in consequence, the intercritically reheated zone of the second cycle does not overlap with the similar zone from the first cycle, as it appears closer to the peak temperature location than in the first cycle (**Figure 29**) [10], while in many true multibead welds, these overlap zones are vulnerable to initiation of voids and cracks, in particular during creep. Result #4 from **Table 3** shows an exceptionally high value due to the overlap of the intercritically reheated HAZ portion of the second simulation cycle (peak 920°C) with the fine martensitic microstructure of the previous simulation cycle (peak 1220°C).

The intercritically reheated zones after a single-cycle simulation at 1220°C appear at about 1.0–1.5 mm closer to the cold Cu jaws, than at the simulation with the peak temperature at 920°C, with the same span between the Cu jaws. To correct this difference, distance plates (P) can be used in the clamping assembly of the Gleeble, and the first thermal cycle, for example, 1220°C, was being executed with both plates inside, while for the second cycle, for example, 920°C peak, one of the plates is removed, and this type of simulation was marked HAZ*. It allowed intercritical zones from both cycles to overlap and resulting microstructures of HAZ* did not differ from those of real welds. To carry out the ACT after the HAZ* cycle simulation, both distance plates have to be removed and thus the free span

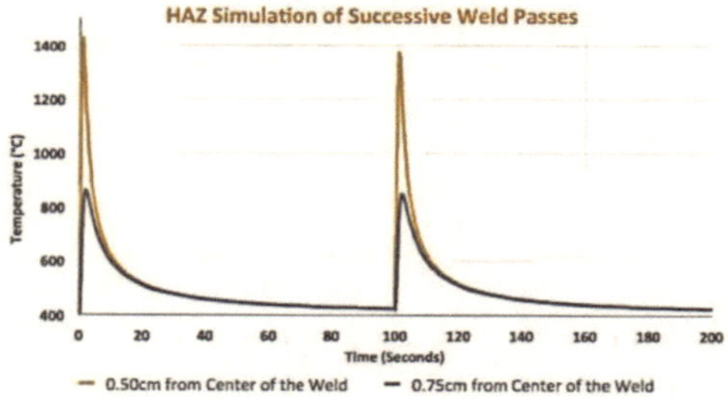

Figure 29.
Thermal graph of two-cycle HAZ simulation on Gleeble [9].

between jaws increased, to secure the uniformly heated hot working zone of about 10 mm length on the ACT sample, as shown in **Figure 30**. The test depicted on this figure was carried out in a low vacuum of the working chamber, so the "hot" portion of the ACT sample visibly oxidized.

The last row (#5) in **Table 3** shows the creep life of two-cycle HAZ*-simulated sample after ACT carried out at lower stress intensity, and the fracture on this sample appeared eccentrically close to the location in which both intercritically reheated zones overlapped. Thin foils for TEM observations taken as close as

Figure 30.
The HAZ-simulated sample in Gleeble chamber during the ACT, before cracking.

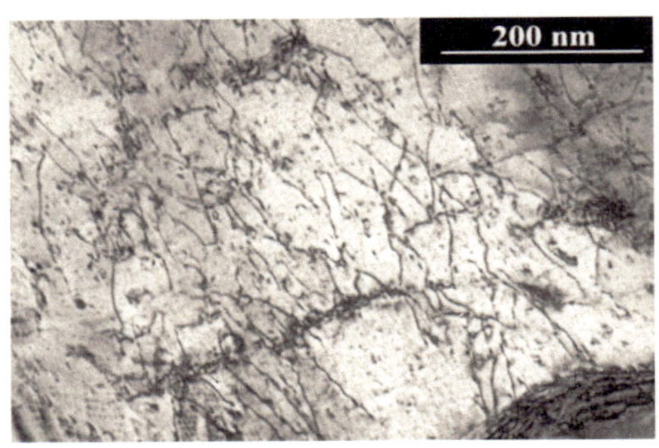

Figure 31.
High density of ultrafine precipitates in ferrite in the two-cycle 1220°C/920°C HAZ-simulated sample (4) with PWHT, after ACT of high stress intensity; foil thickness ~360 nm.

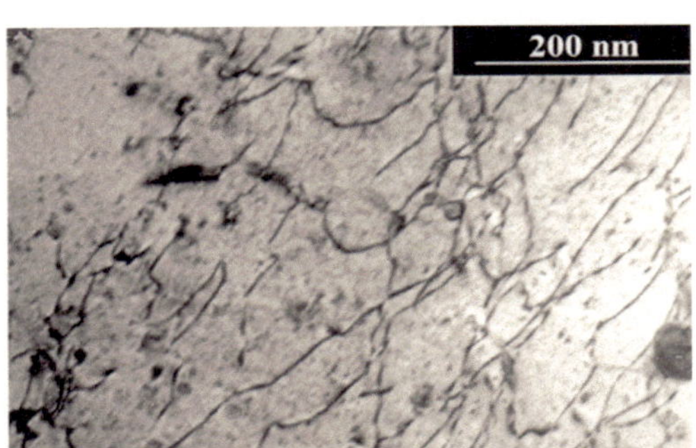

Figure 32.
Dislocations with fine precipitates in ferrite in a single-cycle 1220°C HAZ-simulated sample (3) with PWHT, after ACT of normal stress intensity; foil thickness ~450 nm.

possible (i.e., ~0.3 mm) to the fractures of samples (#4) and (#5) revealed substantial differences of ferrite substructures of the 15Ch2MFA steel HAZs, as to the dislocation density and dispersion of ultrafine precipitates (**Figures 31** and **32**). This strengthening of simulated weld HAZ in the 15Ch2MFA steel as well as increase of its creep life can be related to the strain-induced precipitation forced by multicycle intensive loading during ACT.

6. Evaluation of creep life from ACT data

The up-to-date ACT procedure includes evaluation of true creep life. Out of the collected test's rough data, the stiffness of material, that is, pseudoelasticity modulus E* and then the creep life, can be calculated for a nominal stress, for example, 100 MPa. Some examples from recent research [11], on forged FB2 steel (9Cr-1.5Mo-1.3Co-V-Nb-N-B) and a weld on it, are given in **Table 4** in rows 1–4. In this research, one of the main questions was how much the HAZ on samples taken from real welds has been weaker than the base material, and ACT combined with the HAZ simulation allowed to answer this.

The FB2 grade belongs to the family of new creep resisting steels, developed and tested in EU R&D COST-522 and COST-536 collaborations. In these COSTs, also conventional creep tests were carried out, and for comparison, some of their results are given in **Figure 33**. The result creep life in ACT at 650°C on FB2 steel, that is, 8656 h, matches quite well with the upper black line of the graph presented.

An additional question was if by physical simulation of welding thermal cycles, similar creep properties of HAZ can be obtained and the answer was not straightforward. Physical simulation of two HAZ thermal cycles "one on another," with the first peak temperature of 1220°C followed by the second one of 920°C peak and PWHT "in situ" at 720°C for 15 min, gave in ACT an optimistic result of almost 25,000 h at nominal stress of 100 MPa and temperature of 625°C. The application of the modified HAZ* simulation procedure, mentioned before for the 15Ch2MFA steel, gave after ACT a result similar to creep life of HAZs in real FB2 welds, and the result of this modified HAZ* simulation followed by ACT is included in row 6 of **Table 4**.

The ACT samples for this research were taken from the FB2 material used for the preparation of dissimilar weld joint from forgings of two rings (external diameter 600 mm and thickness 200 mm) made of steel types COST FB 2 and COST F [13]. The heat treatment after the forging of FB2 was 1070°C/6.5 h + 570°C/12.5 h + 710°C/24 h. The weld was manufactured by automated TIG hot wire method in a narrow gap with internal protection by argon, using filler material PSM

	Material	ACT temperature [°C]	Time to fracture [ks]	Average tensile stress [MPa]	Creep life for 100 MPa [h]
1	FB2 base	675	18.04	134	4156
2	FB2 base	650	19.99	149	8656
3	FB2 base	625	9.89	144	27,169
4	FB2 HAZ	625	8.11	196	16,670
5	FB2 HAZ	625	10.59	182	18,636
6	FB2 sim HAZ*	625	9.64	192	17,423

Table 4.
Creep life calculated from ACT data.

Thermanit MTS 616. The thickness of the weld was 120 mm. The chemical compositions of the base material FB2 and the filler metal are given in **Table 5**.

From these welds, smooth cross-weld specimens were fabricated for long-term conventional creep tests (CCT), which were carried out in air at temperatures ranging from 550 to 650°C at stresses from 70 to 220 MPa. Samples from the same weldment were subjected to ACT in order to simulate structure transformations after long-term CCT. ACT was executed at VSB-Technical University Ostrava (Czech Republic) on an HDS-20 physical simulator and partly in the laboratory of the simulator manufacturer (DSI). ACT temperatures ranged from 575 to 650°C (**Figure 33**).

In its initial, as-delivered state, the steel has a high-tempered martensite microstructure, as given in **Figure 13**. This microstructure, as seen on TEM images, shows tempered former martensite laths, retaining high density of dislocations and having chains of particles precipitated along the lath boundaries (**Figure 34**). Most of the precipitates along the former lath boundaries were identified as carbides, while some of the larger-than-average ones appeared to be Laves phase Fe_2Mo. In postmartensitic laths and in elongated ferrite grains, numerous recovery-type planar configurations of criss-crossing dislocations can be observed at larger magnifications (**Figure 35**).

The initial microstructure of the FB2-BM, as seen in light microscope (LM), is highly-tempered martensite of acicular morphology (**Figure 36**). In the as-welded HAZ, it is changed into a very fine-grained one, hard to resolve by light microscopy (**Figure 37**).

Details of the HAZ microstructure are revealed on the following TEM pictures taken from thin foil specimens. As the HAZ is expected to be the weakest part of the weld and consequently of the ACT sample, to take the specimens for TEM observations from exact HAZ location just before their fracturing, the tests had to be run

Material	Element content in wt.pct											
	C	Mn	Si	Cr	Ni	Mo	V	Co	W	Nb	N	B	Al
FB2 steel	0.13	0.34	0.05	9.60	0.17	1.48	0.20	1.32	—	0.059	0.016	0.0079	0.007
MTS 616	0.11	0.32	0.30	8.87	0.57	0.56	0.18	0.15	1.49	0.051	0.018	0.0036	0.009

Table 5.
Chemical compositions of FB2 and weld metal.

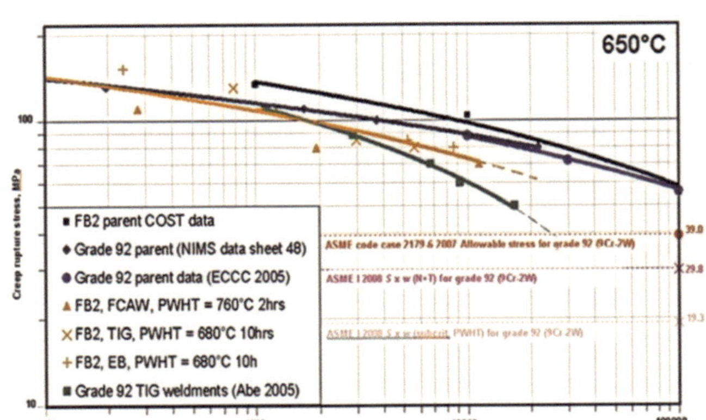

Figure 33.
Creep rupture plot showing the performance of FB2 parent material and weldments, relative to grade 92 (9Cr-2W) seamless pipe [12].

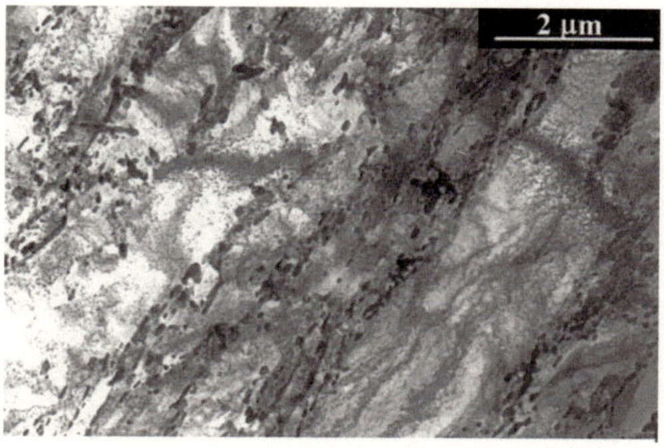

Figure 34.
High-tempered lath martensite microstructure of FB2 steel in as-delivered, i.e., forged and stress-relieved (AD), state.

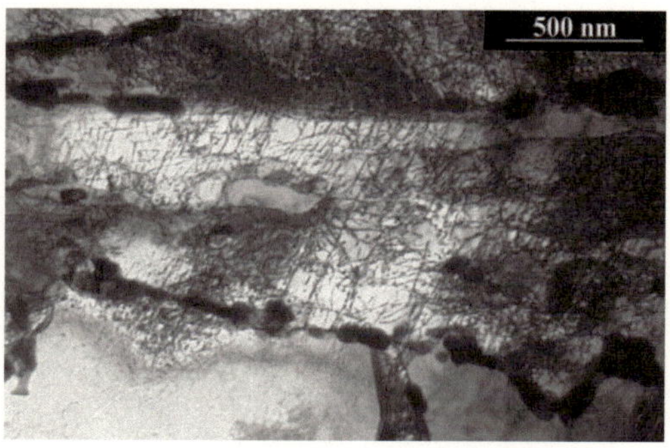

Figure 35.
Criss-crossing dislocation arrays in high-tempered lath martensite in FB2-AD steel with carbides along former lath border.

Figure 36.
FB2-BM microstructure in as-delivered state, before ACT, LM.

till 85–95% of the expected time to fracture. Therefore, in the ACT program, a force limiting stop was placed in the tensile relaxation portion of the testing cycle. Typical force-time graph of ACT on plain FB2 steel (sample FB22) is given in **Figure 38**, showing how the initial tensile relaxation force of ~2200 kgf decreases to ~1400 kgf at 90% of the test duration.

In the initial, as-welded microstructure of HAZ, in its under-critical portion at A1 temperature, the lath martensite recrystallizes, while near to colonies of carbides, first traces of austenite form. Some ferrites get the shape of large laths (**Figure 39**), while in other ferrite grains next to austenite with carbides, very high

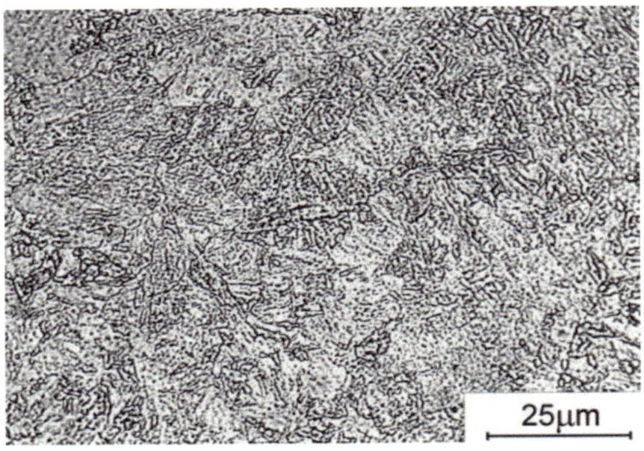

Figure 37.
FB2 weld's HAZ microstructure in as-delivered state, before ACT, LM.

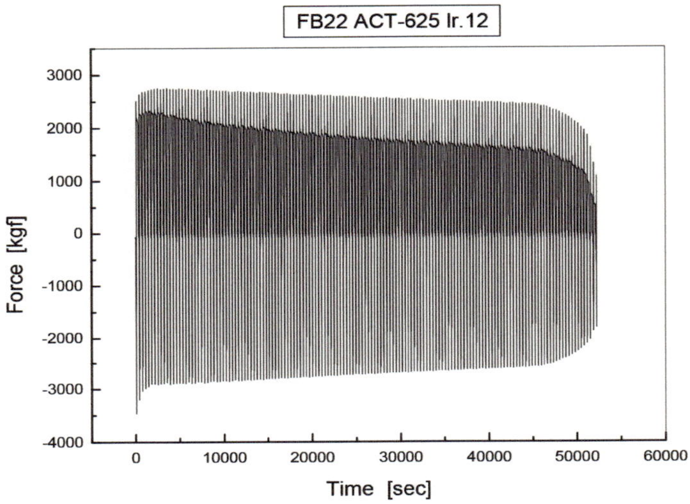

Figure 38.
Typical force-time graph from ACT on FB2 steel.

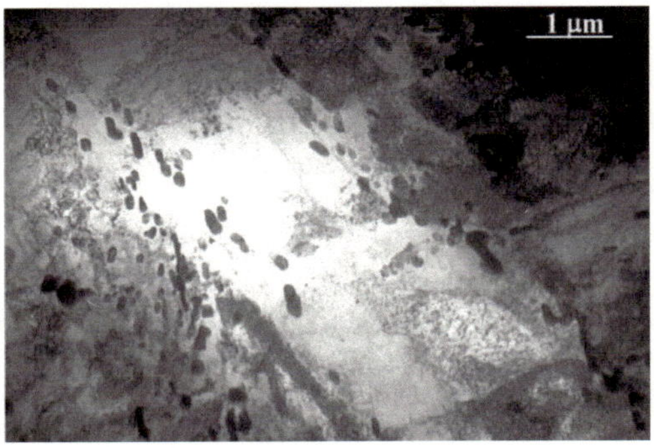

Figure 39.
Large ferrite lath in as-welded FB2 HAZ at location near to A1 temperature.

dislocation density is generated (**Figure 40**). In intercritical portion of HAZ, the austenite forms more compact islands, while some of the carbides still remain undissolved. Finally, above the A3 temperature, austenite becomes more fine-grained homogeneous and close to the fusion surface with the weld metal

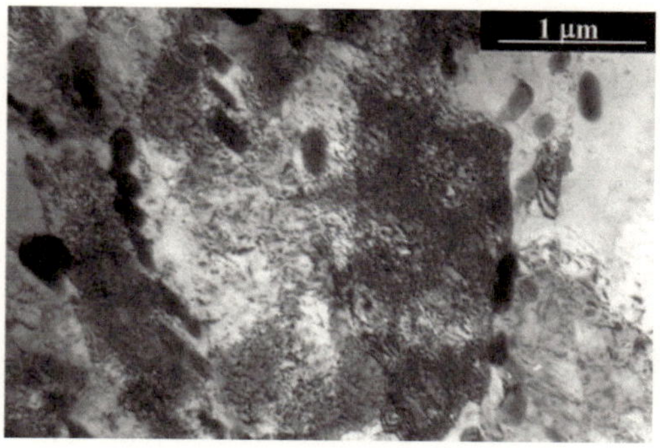

Figure 40.
High dislocation density between nondissolved carbides in as-welded FB2 HAZ microstructure of FB2 in as-delivered state, before ACT, TEM 15,000×.

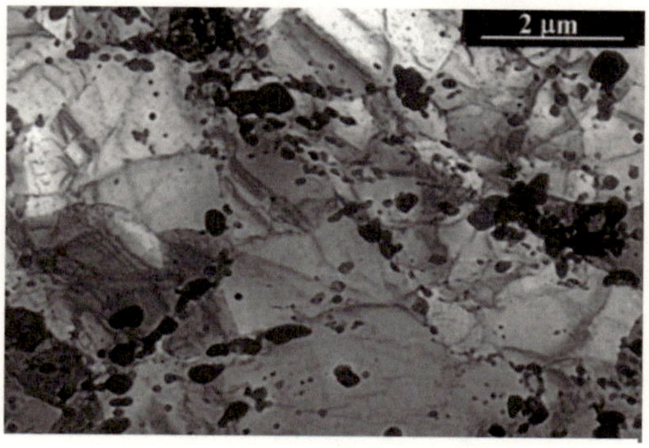

Figure 41.
IC-HAZ in FB2 with recrystallized ferrite and precipitates, after ACT at 625°C.

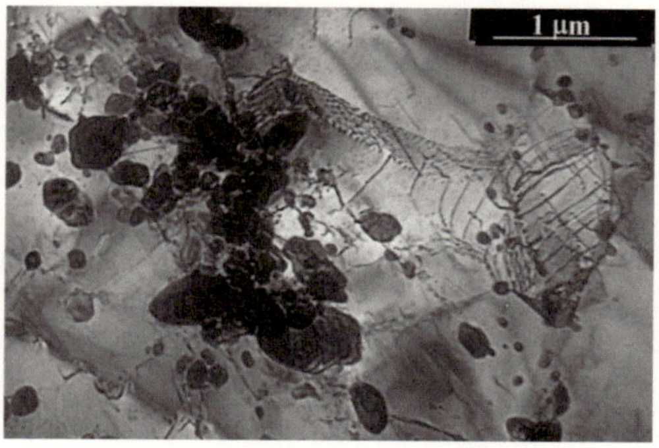

Figure 42.
Agglomerates of precipitates in FB2 weld IC-HAZ after ACT at 625°C.

coarse-grained, which transforms to acicular martensite on cooling. To take the TEM thin foil specimens from the HAZs on ACT samples, from the first tested cross-weld sample, the time to fracture was learned and the force stop equivalent to 90% of this was placed in the program. Applying ACT to the as-welded HAZ microstructures forces their high tempering with very intensive precipitation of carbides. The ferrite in the intercritical portion of HAZ, which contains very few carbides, recrystallizes generating equiaxed grains, while along this ferrite, dense bands with agglomerates of carbides are formed (**Figure 41**). Some largest precipitates in these bands look like undissolved carbides from the initial HAZ

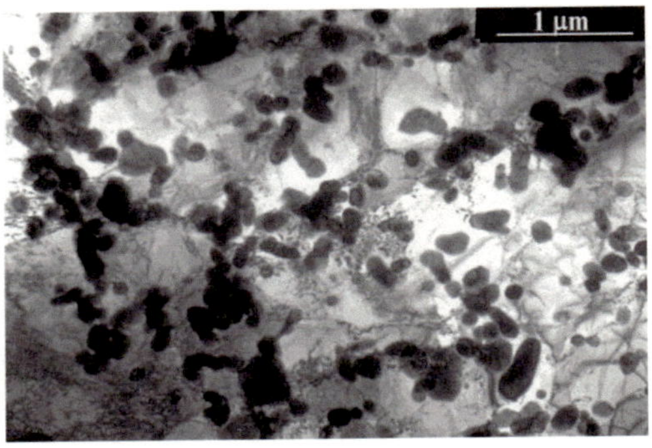

Figure 43.
Carbides in FB2 HAZ at location above A3 temperature, ACT at 625°C.

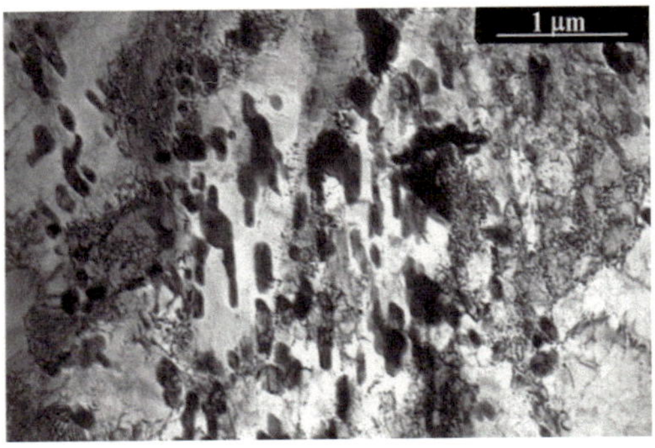

Figure 44.
Carbides in FB2 HAZ at location near to fusion line, ACT at 625°C.

microstructure, while numerous finer ones should be considered as of tempering type (**Figure 42**).

In the HAZ portion representing above A3 temperature, that is, of mainly austenitic as-welded microstructure, high density of partly spheroidized carbides appear and the largest precipitates seldom appear (**Figure 43**), while in the HAZ portion near to the fusion surface, the carbides in high-density colonies are partly elongated and aligned in the directions of former martensite laths (**Figure 44**); here, some larger precipitates do appear again.

ACT executed on the weld HAZ and interrupted before fracturing of sample allows taking proper specimens for TEM observations, which in turn show clearly how the separation of phases, ferrite and austenite, in the intercritical region produces during creep the microstructure vulnerable to premature failure.

7. Discussion

The simulative accelerated creep test was developed for martensitic-ferritic creep resisting steels and welds, based on detailed substructure observations of crept steels. Dislocation configurations governing the real creep were reproduced in a low-cycle compression-tension multicycle procedure carried out at the testing temperatures similar to those of real creep. This procedure applied on Gleeble physical simulator caused in less than 12 h transformation of the initial tempered martensite substructure into fully recrystallized ferrite with spheroidal or coagulated carbides and adequate decrease of hardness similar to that after multiyear exposure to real creep. The carbides and Laves phase precipitates, after ACT, did not reach sizes of these after real creep, but their amount and density appeared substantially larger. The more intensive nucleation of precipitates has been characteristic of electro-thermal treatment [14] and this way the direct resistance heating in Gleeble [1] plays a role in acceleration of the process. The larger density of precipitates, despite their smaller sizes, results in a similar or even more substantial depletion of matrix in alloying elements like in the long-term real creep. The strains applied in each cycle of the ACT procedure are too small to homogenously deform the tested material in any of the individual cycles to cause dynamic recovery or recrystallization. The dislocation configurations generated in each deformation cycle accumulate from cycle to cycle mainly in the preferentially oriented components of microstructure and after reaching critical density annihilate, gradually contributing to generation of voids and cracks, simulating the real creep in this way. Finally, on the crack surfaces after ACT, very large amounts of precipitates are present and also characteristic lines of dislocation escape appear. As to the chemical compositions of precipitates after ACT, in an earlier study on P91 steel [15], the microanalysis results from precipitates after ACT matched those of the crept steels and complied with ThermoCalc equilibrium predictions.

Up to now, the ACT procedure was applied in a study on development of P24 grade welding consumables [16] and new steels for supercritical components of power generation systems [17], as well as for determining remnant life of crept steels and repair welds [18]. Finally, in combination with weld thermal cycle simulation, it offered an opportunity of fast optimization for welding procedures to address problems of weakening heat-affected zones of welds.

The ACT procedure does not force the material to fail by overloading. By cyclic compressing and stretching of the tested material, ACT records response of the material while choosing the most prone sites in microstructure for nucleation of voids and cracks, thus well simulating the real creep. Programmed various amounts of compressive and tensile strains in combination with thermal cycles affect the intensity and duration of test as well as promote precipitation, making in intended manner the ACT plausible for study of creep-fatigue situations, importance of the last has been growing recently due to possible instability of grid by incorporating various energy sources. The conventional standard long-term creep tests cannot address this last issue.

8. Conclusions

1. Low-cycle thermal-mechanical fatigue of martensitic-ferritic creep resisting steels, applying direct electric resistance heating, accelerates transformation of microstructure and intensifies precipitation of secondary phases, causing depletion of matrix in alloying elements similar to that occurring during the real creep.

2. A substantial role in the accelerated microstructure transformation is played by meta-stable a<001> edge dislocations, which effectively transport interstitials, thus accelerating precipitation.

3. From the accelerated creep tests as well as from conventional case studies appears that the applied stresses and accumulated strains speed-up weakening of the matrix and promote local failures in the most vulnerable sites of microstructure.

4. Hardness measured at room temperature after ACT indicates the extent of microstructure transformation; however, when the material will fail cannot be predicted.

5. Calculation of true creep life from ACT is possible, taking into account changing of stiffness of material at the elevated temperature of the test, its flow stress, and duration of the test.

6. A combination of the weld's HAZ physical simulation with the direct execution of ACT allows speeding up the development of optimum welding technology.

7. In general, within the relatively short time of the ACT's duration, large amount of data can be generated, which are needed for the development of improved new materials and technologies.

8. The simulative accelerated creep test can be a useful alternative to the existing standardized long-term tests especially that it is giving adequate transformation of microstructure in much shorter time than the conventional tests and can also provide materials data relevant to creep-fatigue situations.

Author details

Stan T. Mandziej
Advanced Materials Analysis, Enschede, The Netherlands

*Address all correspondence to: stanamanl@cs.com

References

[1] Available from: www.gleeble.com/products.html

[2] Honeycombe RWK. Creep in pure metals and alloys. In: The Plastic Deformation of Metals. London: E. Arnold Publishers; 1984. p. 356

[3] Sleeswyk AW, Mandziej S. Mobility of <001> dislocations and formation of cleavage nuclei. Scripta Metallurgica et Materialia. 1990;**24**(1):7

[4] Výrostková A. IMR-SAS, Kosice SK. Case study in COST-538 Action. 2005. Unpublished. Available from: http://www.cost.eu/domains_actions/mpns/Actions/538

[5] Stroh AN. A theory of the fracture of metals. Advances in Physics. 1957;**6**:418

[6] Mandziej ST. Low-energy dislocations and ductility of ferritic steels. Materials Science & Engineering A. 1993;**164**:275

[7] Sleeswyk AW, Mandziej S. Dislocation behaviour in dual-phase steel. In: Kettunen PO et al, editors. Strength of Metals and Alloys. Vol. III. Oxford, UK: Pergamon Press; 1988. p. 155

[8] Lubahn JD, Felgar RP. Plasticity and Creep of Metals. New York: J. Wiley & Sons Inc; 1961

[9] Mandziej ST. Microstructures of heat-affected zones in Cr-Mo-V steels; IIW Doc. II-C-514-17

[10] Gleeble Welding Simulation Brochure 6.12. Poestenkill, USA: DSI; 2016

[11] Mandziej ST. Accelerated creep test for new steels and welds. In: Liu X et al, editors. Energy Materials 2017, The Minerals, Metals & Materials Series. Springer Verlag; 2017. p. 161

[12] Rothwell J, Abson D. Performance of weldments in advanced 9%Cr steel—'FB2'. Materials at High Temperatures. 2010;**27**:253

[13] Kasl J, Jandova D, Kanta V. Behaviour of dissimilar weld joint of steels FB2 and F during long-term creep test. IOP Conference Series: Materials Science and Engineering. 2019;**461**:012036

[14] Kidin IN. 'Kinetics of Thermal Processes at Different Heating Methods', on p. 42 and 'Electrothermal Treatment of Alloy Steels', on p. 191. In: Fizicheskie Osnovy Elektrotermicheskei Obrabotki Metallov i Splavov (Physical Principles of Electrothermal Treatment of Metals and Alloys). Moscow: Izd. Metallurgia; 1969 (in Russian)

[15] Mandziej ST, Vyrostkova A. Evolution of Cr-Mo-V weld metal microstructure during creep testing—Part 1: P91 material. Welding in the World. 2008;**52**(1–2):3-26

[16] Mandziej ST, Vyrostkova A, Solar M: Evolution of Cr-Mo-V weld metal microstructure during creep testing — Part 2: P24 material. Welding in the World. May 2011;**55**(5-6):52-69

[17] Mandziej ST. ACT study in COST-536 Action. 2005. Unpublished. Available from: http://www.cost.eu/domains actions/mpns/Actions/536

[18] Mandziej ST. Remnant life estimation for repair welding in thermal power generation. In: Design, Fabrication and Economy of Metal Structures, International Conference Proceedings, Miskolc. Hungary: Springer; 2013. p. 263